TRACK DESIGN 2

Edited by Hal Carstens

ISBN 911868-62-3

Dedicated to William Schopp who for many years created the Layout Doctor series in Railroad Model Craftsman.

Carstens
PUBLICATIONS, INC.

PO BOX 700
NEWTON, NEW JERSEY 07860

Contents

COVER: An N&W GP38AC on a local crosses over a departing freight led by a Susquehanna F45 on the HO scale Rochester Model Railroad Club. Massive bridges and mountains are basic ploys in making most model railroads seem larger than they are, in addition to setting the stage for train operation. The diesel modeling and Kodachrome cover are the work of William C. Baker.

An Introduction to Track Design

Wide sweeping curves with gentle grades on Hal Carstens' Susquehanna Northern Railroad were easy to create with spline subroadbed. Before construction began, the background scenery was painted with acrylics to form mountains and sky. All the risers were fastened with white glue clamped in place until dry.

The designing and creation of a new model railroad layout requires lots of thought, planning and dreaming. That dream railroad you want to build will be an extension of yourself. It will represent your idea of the ideal model railroad adapted to whatever restrictions you may impose. You may opt for Lionel track, sectional track, flexible track, or track laid by hand on wooden ties. You can model prairies, mountains, coastal areas, cities, industrial areas or mining. You can model any period in railroad history you like, from the pioneer days of 1830 to heavy diesel operations of today, with Amtrak, VIA and commuter operations thrown in as an extra. You can fill an entire basement or garage or you can restrict yourself to a narrow shelf in a bedroom. Model railroading knows no limits.

If you have a lot of equipment, economics might dictate that you stay in that scale and gauge. Certainly the time to make a change is before you start designing the new track plan. Many modelers have traded off their old equipment for something in a new scale. An HOer with loads of ready to run equipment might decide to trade it off for one On3 brass locomotive. An O gauger moving to Florida might unload his collection in favor of a smaller gauge which would take up less room in a smaller home. Some guys go to a larger scale because their eyes aren't as sharp as they used to be.

What's the difference between scale and gauge? Many modelers use the terms interchangeably but remember that gauge refers to the width of the track between running rails, while scale is the scale in which you're modeling. An HO narrow gauger could well be using equipment that would run on N scale track. A narrow gauge modeler in O scale would not be running on O gauge track, although his modeling is in the same proportion as the modeler modeling Santa Fe or Conrail. G gauge trains may be narrow gauge, or standard gauge made to several different size standards and they'll all run on the same width track.

Space Requirements

Once you've chosen a scale, you have to consider the space available for your layout. In many cases the modeler will have to be content with a 4x8-foot pike, which means that the maximum curves will very likely be 18" or 20" radii. S and O gaugers faced with this dilemma are almost forced to go the tinplate route unless they're in traction. Eighty-foot passenger cars in HO scale can be made to run on 24" radius curves, but they'll run better on 30" or 36" curves, and they'll look better on 48" curves because the unsightly overhang will be avoided.

If you must have sharp curves, limit your rolling stock to a maximum of 60 or 70 feet in length. There are passenger cars available as short as 40 ft. and even 36 ft. in length. Trains will seem longer because you can operate more cars in the same lineal space and the shorter equipment will look more natural on the sharp curves. Many HO locomotives, especially brass steam locomotives, require curves of 30" or larger. Some O scale locomotives may require radius of 60" or larger. And remember that some 4-4-0 types may require a wider operating radius than certain larger wheelbase locomotives because of the rigidity of their frames. A good hobby dealer can answer many of your questions for you. **Paul Mallery's Design Handbook for Model Railroads** available from your dealer can answer many design questions that will eliminate problems as you proceed with construction.

Where the modeler's apartment is small, the layout may have to be crammed into a 3'x4' or even a 2'x3' space. Don't laugh. Many very fine model layouts have been built in such limited space but it is almost mandatory that you restrict yourself to HO, N, or Z scale. In HO you'll want to use the smallest locos and cars not over 40 feet long.

Another alternative to building a layout in a restricted space is to build a module, perhaps 2'x6' or 2'x8'. Operation will be limited to switching, which many modelers feel is the ultimate reason for having a layout anyway. Make sure your module adheres to standards used up by the local modular groups so you can connect your module with others at conventions and train meets where you'll be able to run freight trains of 200 cars in length over your little section!

What Kind of Railroad To Model?

You'll want to consider what type of railroad you want. Will you model the Pennsylvania Railroad's four-track mainline, or a western mainline with only one track such as the Santa Fe? Many railroads have eliminated multi-track mainlines in recent years. Would you prefer a medium-sized or short line operation? A steam era railroad will require turntables and other facilities not needed on a modern diesel-only CSX layout, although there well could be a roundhouse and turntable on the latter.

Narrow gauge modeling in two foot, three foot or meter gauge holds a fascination for many modelers. So does trolley modeling and there are trolleys available for most of the bigger systems and many smaller ones as well. Other modelers find European modeling to their liking and hobby shops carry a good assortment of equipment for British, German, French and other European railroads.

Loop or Point to Point?

Will your layout be point-to-point or would you prefer a loop? Some modelers like the point-to-point system, feeling it is more prototypical even though it will require constant making and breaking of trains as a run is finished. Modelers who prefer operation often opt for this type of layout and they'll often have lots of switching in yards and industries.

The loop enthusiasts, which means any track design having continuous operation, enjoy long running and will often store a train as is at the end of a run and then pull out another. Some layouts, such as my Susquehanna Northern, are point-to-point but have hidden loops built in at each end so that either type of operation can be enjoyed. Many smaller layouts are loops because they offer longer runs.

A variation of the point to point layout is the out and home layout having only one stub and terminal and a loop at the other end. Dogbone layouts are loops having the connecting straight tracks close together as though double tracked. The track kit found in Complete Layout Plans lets you plan many types of layouts even though the track segments use HO sectional track pieces. Figure out your basic design, then transfer it to your special size needs, preferred radius, and other preferred requirements. Remember that you can keep making changes until the track is laid. Changes can also be made after your track is laid but it's best to plan carefully before you start track laying.

Potential Problems to Solve

There are things you'll want to consider in designing your layout. For best operation, track grades shouldn't exceed 2% and yet many model pikes have grades of 4% and more. Tablework at about 40" suits many modelers, but others prefer a higher "water" level, while some modelers like a layout down low.

Funny things can happen if you don't pay attention to potential snags. An around the walls plan for N or HO could be enlarged for use in O scale, but be careful that your span of reach doesn't spread to 5 or 6 feet without adding access holes where needed! Sometimes a removable section will solve an access problem but keep it simple. Duck unders are useful, but as the late Bill Schopp said, "They became progressively more difficult to negotiate as one grows older". My Susquehanna Northern avoided duck unders but I still occasionally have to crawl under to rerail a car, fix wiring or make some repair or adjustment.

Walk Around Layouts

Where space permits, walk around cab control lets the modeler/engineer follow the train with a controller in his hand as he walks around the layout. Walk around control is at its best where tracks are designed to be fairly close to the aisles allowing the engineer to walk along with his train. A greater concept of depth and distance is created by designing the layout so that tracks on the far side of the aisle are hidden by mountains or by a painted mountain scenery section. Each scenery segment could have a different scenic treatment, even seasonal. Walk around layouts are excellent for use with the several cab control systems available which simplify wiring.

Designing the Layout

Where do you get ideas for the layout? Start by browsing through back issues of RAILROAD MODEL CRAFTSMAN to see what others guys did. You'll soon find a layout that comes close to what you want. Also see your hobby dealer for one of the track books he has in stock, such as **Complete Layout Plans, Design Handbook for Model Railroaders,** or **McClellands's Virginian & Ohio.** When starting your planning, don't worry whether a layout design uses toy, sectional, or flexible track. Design that first route and later you can figure out what type of track you want to use.

If you're relatively new at the hobby, try not to bite off more than you can chew. A good layout can be simple and relatively small. Many modelers have designed super pikes beyond the modeler's capacity to complete in time, talent and money. Build a small layout first. You can always add to it after it is completed. That's the way John Allen started his great Gorre & Daphetid Railroad.

It may be that you will throw up your hands at designing a layout, opting to have a track design expert do the work for you. Among such design experts is Murray Bloom who is a whiz at designing layout plans for others. For roughly the price of a really good HO brass locomotive (the price depends on the size and complexity of the layout), Murray will create a custom design. He first sends a questionnaire, which gives him an idea of the type of layout you have in mind: time, period, geographic location and state, terrain, landscape, railroad type, prototype if any, and he'll want to know if you want point-to-point, loop-to-loop, out-and-back or some other alternative. He also asks about scale and gauge preferences, electrical preferences, turnouts and other specifics.

Don't expect a specialist like Murray to return your layout in the next day's mail. After a waiting list period, expect about one or two months for a preliminary design, another month or two for track plans and the final Railplan in another one or two months. Murray has done over 500 such plans in almost every scale, gauge and size. He'll send a complete information and start-up package ($3 refundable) by writing Railplan, 1269 Vogt Ave., Baltimore, MD 21227. You'll find ads for track designers in RAILROAD MODEL CRAFTSMAN.

NMRA Standards

When you build a railroad, it is essential that yhou adhere to certain standards of construction so that your layout will be able to run the equipment you and visiting modelers will want to run. Almost all model railroad equipment sold in north America follows NMRAs standards. European equipment is very similar but wheel flanges are often somewhat larger, operating best on code 100 rail in HO, causing problems when operating on code 83 or code 70 rail in HO. Märklin HO uses inside third rail or stud contact. Be sure you understand that the nice British equipment you bought in England may not run on your layout without some work on the wheels to make them compatible.

Track Choices

Today's modeler has a wide choice of track materials from which to choose. Still popular and easy to use are the tubular track systems used by Lionel, American Flyer and K-Line trains. G gauge uses large solid T rails mounted on heavy plastic ties. Atlas offers O scale track sections mounted on plastic ties. Similar track is offered in all the other popullar scales (HO, Z, N, O). If your layout is housed in a room having a moisture problem avoid rails prone to rust or oxidation. Many modelers prefer nickel silver rail which requires a minimum of cleaning. Beginners will find it easier to use switches which have the switch machines built in, eliminating the hook up problems at least on that first layout.

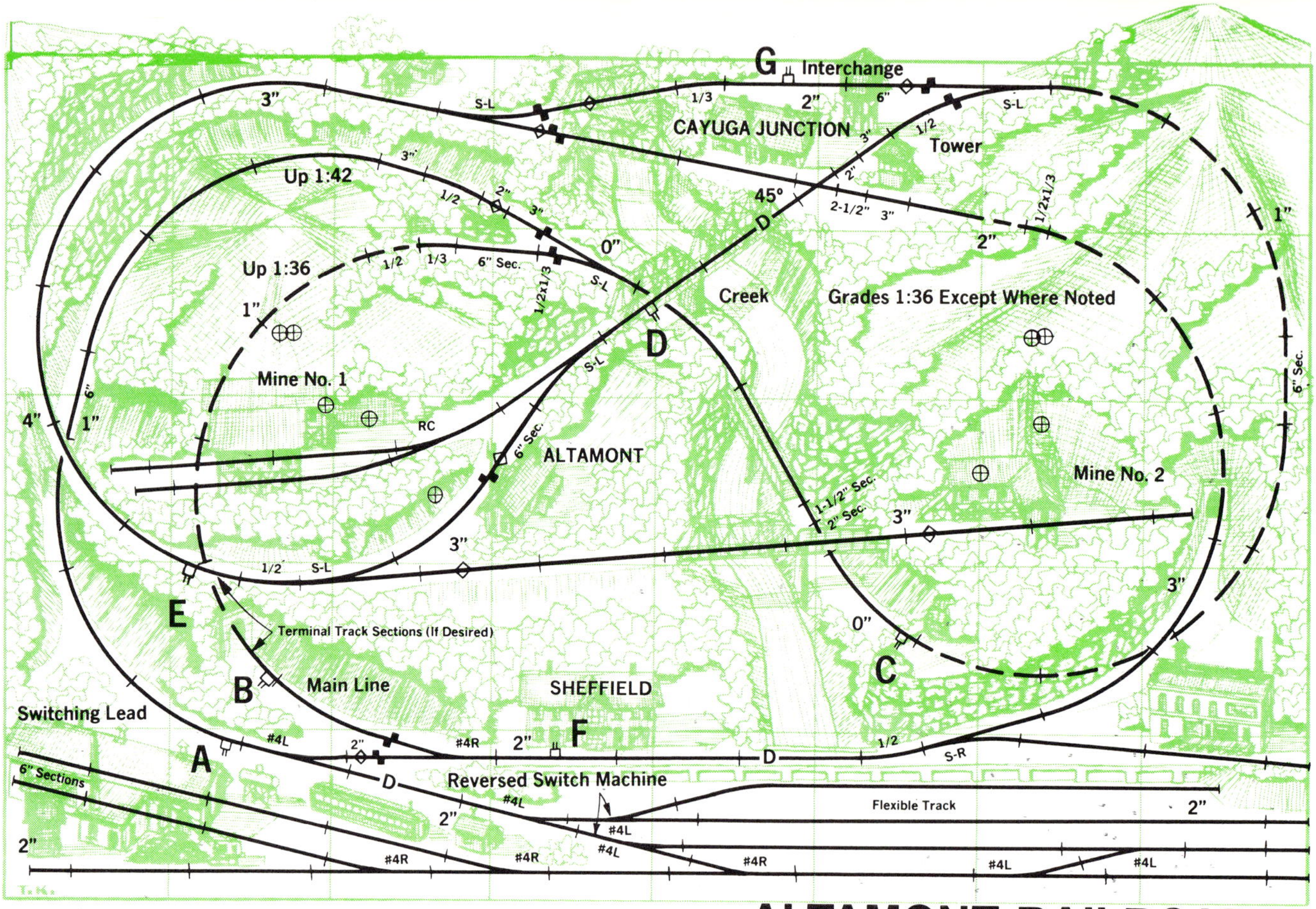

ALTAMONT RAILROAD

Scenery and Operation in 5x8

Whether sectional, flexible or hand laid track is used, the scenicking and operating possibilities on this HO pike are suited for beginners and veterans alike.

For many beginning model railroaders, the first step in constructing their first layout is to head for the local lumber yard for a sheet of 5/8" or 3/4" plywood, usually 4'x8' in size. Trackplan books offer several worthwhile sectional trackplans specifically designed for the Good Old 4x8, and the beginner most often selects a plan which offers a loop (perhaps cleverly disguised by the track planner) for continuous running, plus a few turnouts (switches) to add interest.

Fortunately, most published plans include a small yard, a passing siding and hopefully a few industrial sidings to allow the modeler to run his trains with at least some degree of prototypical realism in operation after the novelty of running the trains around the loop has worn off. Beginners who fail to include the "secondary" trackage necessary for operation of one type or another often lose interest after a rather short period of time—and perhaps drop out of the hobby altogether without ever having gained a real understanding of the true potential for enjoyment offered by model railroading.

The continued growth of the model railroading industry (and RMC's circulation) indicates that a good many of the beginners stay on long enough to grasp what railroading—and hence *model* railroading—is all about. With this increased knowledge comes an awareness that the Good Old 4x8 pike, usually built on one level, has some definite limitations, not only from an operational standpoint, but scenically as well.

Too often the blame is placed on the 4x8 ("better to build an around-the-wall layout") or on the sectional track which came with the trainset ("better get rid of that sectional track; it restricts your ability to add realism to the trackplan too much"). In fact, neither of these factors are usually to blame for the failure of a given 4x8 sectional pike to maintain a high interest level on the part of the beginning model railroader.

To illustrate the point, RMC's editors designed a layout which would include most of the features necessary to maintain a high degree of interest for most modelers due to its operational and scenic possibilities—and we designed it to fit approximately in the space now occupied by that Good Old 4x8 (it actually measures 5x8). It is also designed to utilize available sectional track components except in a few special cases where the three foot lengths of flexible track are needed. Flexible track must have the inner rail cut to length after it is curved to the desired radius, but an inexpensive razor knife such as the one made by X-acto handles this task nicely. A motor tool like Dremel's is ideal for the purpose, and is an excellent buy even for the rank beginner due to its many home and hobby applications.

The original trackplan was sketched on a sheet of paper in the time honored tradition. It is dangerous to go beyond the rough outline of the mainline at this stage, however, since it is far too easy to tighten up a curve here and there to get in an extra yard track. Then, when the design is drawn to scale, either a number of tracks must be eliminated to maintain reasonably large mainline radii, or too many tracks are jammed into the remaining space in an attempt to retain the operation features, sacrificing scenic realism.

Fortunately, the original sketch was reasonably accurate, and the Layout Doctor

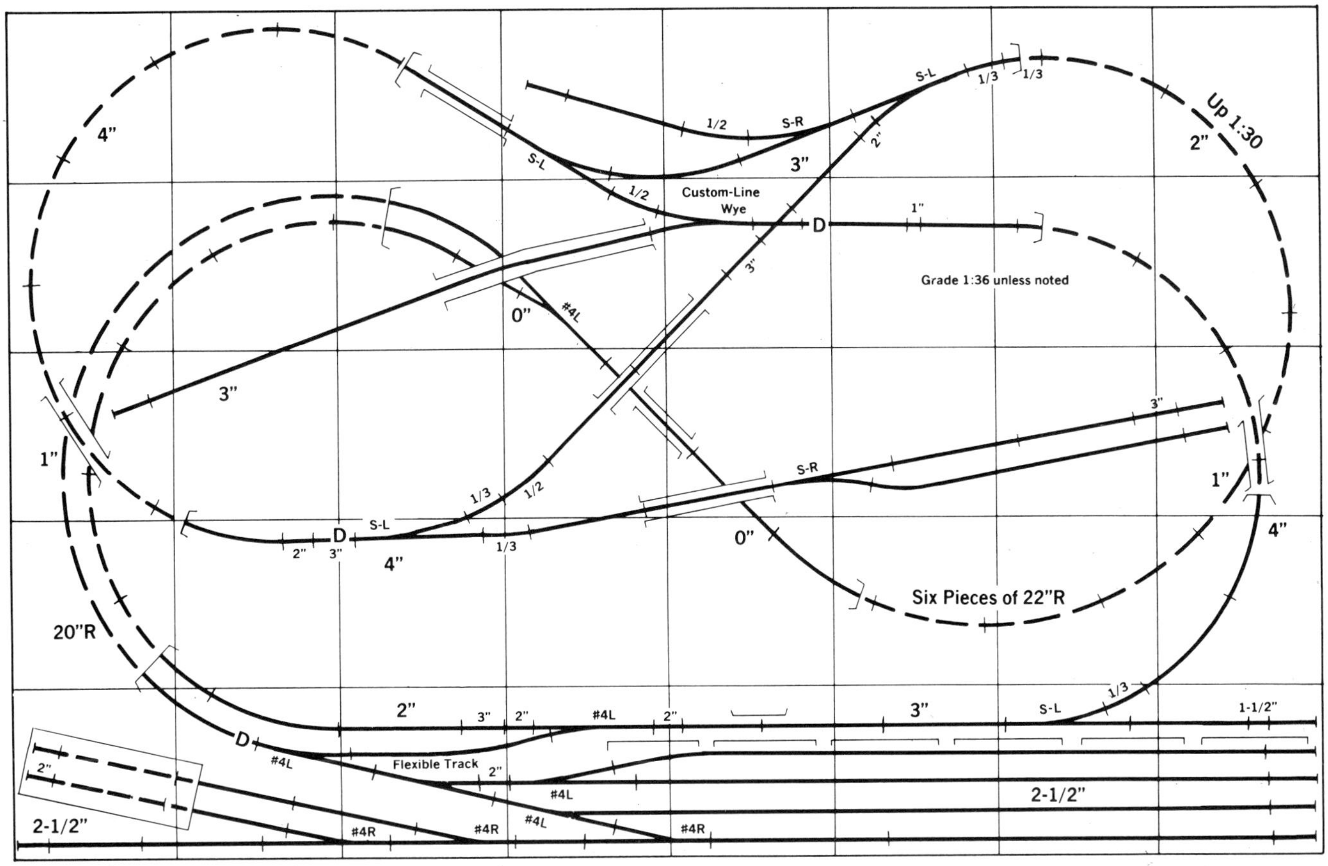

Trackplan Modifications

The original design is indicated above; comparison with the revised design on page 6 shows the "uncovered" yard lead and revisions in track location near Altamont and Cayuga Junction. A third design combining the best features of both would be easy to work up.

Layout Symbols

Size of squares: 12" for HO Scale

Power Feeders	A Block Designation
Insulated Rail Joiners (Gaps)	0" Track Elevation
—D— Delayed Magnetic Uncoupler	Curve Radius 18"
Magnetic Uncoupler	⊕ Center of Major Curves

was able to make a scale drawing of the plan with very few changes due to space limitations. This is shown on page 35.

However, a careful study of that design revealed potential problems, especially for the beginner, so this design was modified to the "final" design shown with scenery on page 34. A variation using a separate leg for a yard was also prepared for those with extra space (page 36). We will discuss the reasons behind the original design, plus those behind the modifications, so that beginners can understand why a great deal of thought must go into the initial design stages of any proposed layout to avoid problems later on when design revisions are difficult or impossible.

A multi-level plan was desired to allow for increased mainline length, a feature found on most trackplans. Since a turntable is somewhat space consuming, a turning wye for train and locomotive reversing was desired; a bit of planning allowed us to include the wye on the main rather than in the yard to set up various routes for trains, simulating branch lines or even junctions with other railroads. The wye is at Cayuga Junction; trace the path of a westbound train leaving Sheffield yard, then through the right-hand route of the turnout (avoiding the 45° crossing) and you'll find the train will arrive heading eastbound at Sheffield. An eastbound movement from Sheffield must back through the wye if a westbound arrival at Sheffield is desired, but this time-consuming movement is ideal for modeling purposes (remember that *time is distance;* the longer it takes to make a circuit, the longer the apparent length of the railroad).

Another feature of the original and modified designs deals with mine runs. Obviously, it is easy to switch Mine No. 1 by heading west out of Sheffield, through Cayuga Jct. over the 45° crossing, past Altamont, again over the crossing at Cayuga Jct. on the other main—and finally backing into the mine spur at the Junction. Mine No. 2 is even easier to switch for a westbound out of Sheffield.

Now comes the fun: an eastbound leaving Sheffield can switch Mine No. 2 by backing through the wye ("interchange") track at Cayuga Jct., then pulling ahead to Altamont. Thus, Mine No. 2 may be considered Mine No. 3 for eastbound mine runs, effectively adding an industry to the railroad without consuming more space. Mine No. 1 similarly becomes Mine No. 4 for eastbounds. By backing through the wye at Cayuga Jct., these mine runs can then return to Sheffield yard as eastbounds, or they can continue along as westbound trains after the mine work is completed. Obviously, this design is not limited in scope to the beginner; a veteran modeler could devise countless variations for mine runs, local passenger trains taking workers to the mines (using gas-electrics, McKeen cars, RDC's or a small steamer and some ancient varnish) and so on. It could easily become a dispatching nightmare (which is desirable!), but the beginner need not involve himself to that degree in operation to enjoy this railroad.

The scenery as drawn is just one possibility; the coal mines could be changed to industries more typical of some parts of the West, for example.

By treating the mine runs as *branchline* operations rather than mainline movements, the railroad is simple in actuality but extremely complex in theory: it then becomes a mainline heading both east and west out of Sheffield, with several branch lines also reaching into the mountains for coal out of the same point. Cayuga Junction could be assigned several names, depending what type of train was heading through it (main or branch). Other modifications

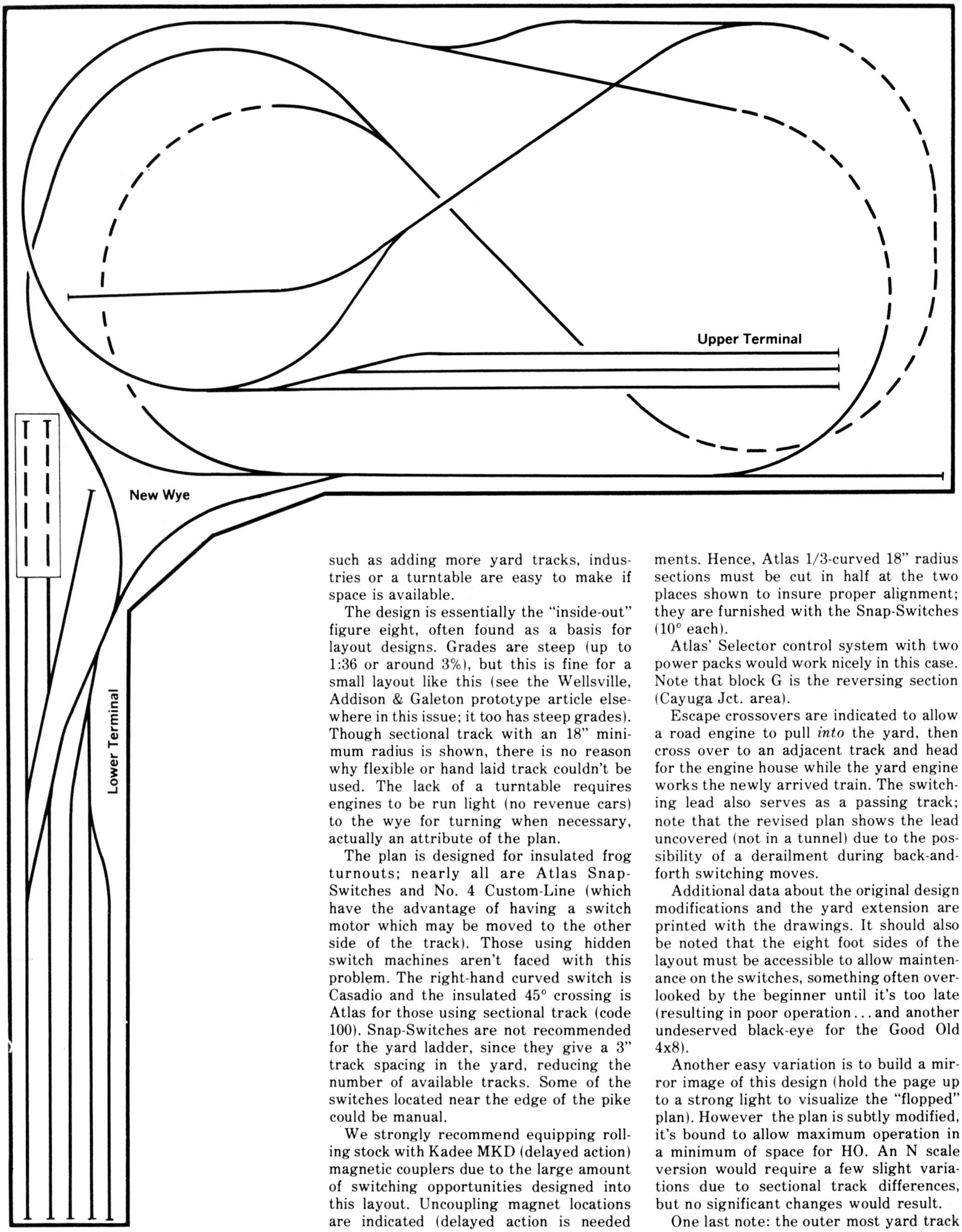

Adding A Yard

The trackplan above is basically the same as that shown on page 5, though the yard has been moved to a separate narrow shelf. A wye at the yard throat has also been added which can be used to turn locomotives or entire trains.

such as adding more yard tracks, industries or a turntable are easy to make if space is available.

The design is essentially the "inside-out" figure eight, often found as a basis for layout designs. Grades are steep (up to 1:36 or around 3%), but this is fine for a small layout like this (see the Wellsville, Addison & Galeton prototype article elsewhere in this issue; it too has steep grades). Though sectional track with an 18" minimum radius is shown, there is no reason why flexible or hand laid track couldn't be used. The lack of a turntable requires engines to be run light (no revenue cars) to the wye for turning when necessary, actually an attribute of the plan.

The plan is designed for insulated frog turnouts; nearly all are Atlas Snap-Switches and No. 4 Custom-Line (which have the advantage of having a switch motor which may be moved to the other side of the track). Those using hidden switch machines aren't faced with this problem. The right-hand curved switch is Casadio and the insulated 45° crossing is Atlas for those using sectional track (code 100). Snap-Switches are not recommended for the yard ladder, since they give a 3" track spacing in the yard, reducing the number of available tracks. Some of the switches located near the edge of the pike could be manual.

We strongly recommend equipping rolling stock with Kadee MKD (delayed action) magnetic couplers due to the large amount of switching opportunities designed into this layout. Uncoupling magnet locations are indicated (delayed action is needed at locations marked "D"); some revision may be necessary.

In studying various Atlas layout books, it was noted that arcs of 2-1/2° are at times "forgotten" when laying out curves, and no problems are encountered as far as we can determine; sectional track can be "forced" slightly to allow for small alignment errors. However, 5° arc errors would almost certainly cause a kink, and subsequent derailments. Hence, Atlas 1/3-curved 18" radius sections must be cut in half at the two places shown to insure proper alignment; they are furnished with the Snap-Switches (10° each).

Atlas' Selector control system with two power packs would work nicely in this case. Note that block G is the reversing section (Cayuga Jct. area).

Escape crossovers are indicated to allow a road engine to pull *into* the yard, then cross over to an adjacent track and head for the engine house while the yard engine works the newly arrived train. The switching lead also serves as a passing track; note that the revised plan shows the lead uncovered (not in a tunnel) due to the possibility of a derailment during back-and-forth switching moves.

Additional data about the original design modifications and the yard extension are printed with the drawings. It should also be noted that the eight foot sides of the layout must be accessible to allow maintenance on the switches, something often overlooked by the beginner until it's too late (resulting in poor operation . . . and another undeserved black-eye for the Good Old 4x8).

Another easy variation is to build a mirror image of this design (hold the page up to a strong light to visualize the "flopped" plan). However the plan is subtly modified, it's bound to allow maximum operation in a minimum of space for HO. An N scale version would require a few slight variations due to sectional track differences, but no significant changes would result.

One last note: the outer most yard track could be made removable. A cut of cars would be shoved onto this section, which would then be lifted out and stored on a shelf. Another empty or already loaded track section would be slipped in place, representing an interchange track with another railroad or belt line. John Allen used a car ferry on his G&D Line (another car ferry design is shown on page 40) to accomplish the same purpose.

by Bill Livingston

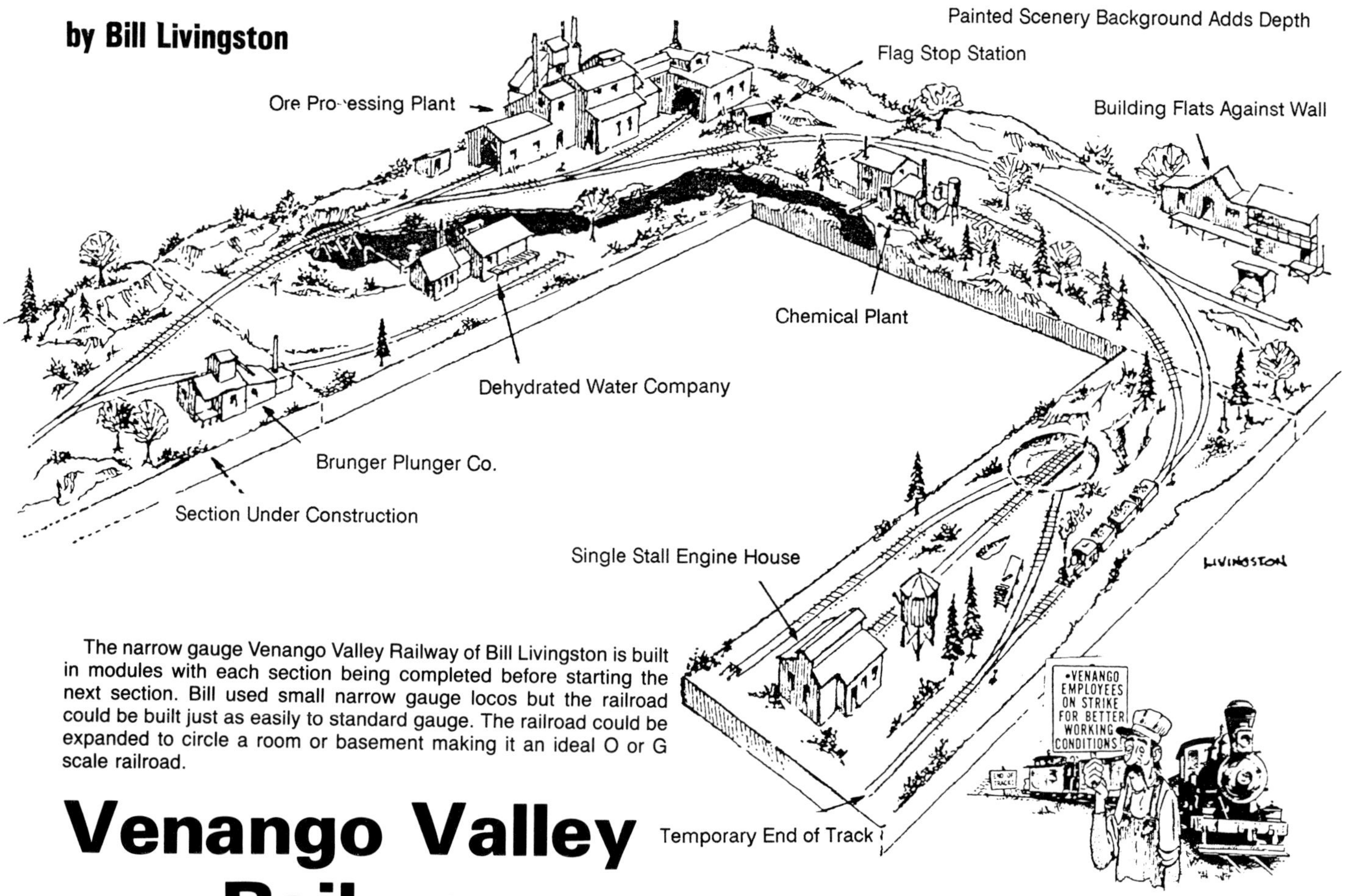

The narrow gauge Venango Valley Railway of Bill Livingston is built in modules with each section being completed before starting the next section. Bill used small narrow gauge locos but the railroad could be built just as easily to standard gauge. The railroad could be expanded to circle a room or basement making it an ideal O or G scale railroad.

Venango Valley Railway

Modules were built 18" wide x 54" long, carriage bolted to gether and bracketed to the wall at near eye-level.

Hobby shops carry flexible track in yard long strips in which T rails are mounted on plastic tie strip. You can also buy rail stock in brass or nickel silver which you must spike to real wood ties using tiny spikes or wire staples using a rail spiker such as the one offered by Kadee.

Whatever your choice of track, it is advisable to mount the track on roadbed. Roadbed made of rubber, cork, or Homosote tends to deaden the sound of the trains, especially where plywood is used for the table.

Rules for Good Operation

We've mentioned that wide curves can provide better operation than sharp curves. Where sharp curves are necessary, keep your train speeds under control. Real trains don't do 200 mph around hairpin turns. Try to keep train speeds at real train speeds depending on your type of railroad. A model train doing a scale 60 miles per hour would take one minute to cover a scale mile regardless of scale or two minutes at 30 miles per hour. The slower the train, the longer the run in time. A mile in some of the popular model train scales are:

Z Gauge	24 ft.
N gauge	33 ft.
HO gauge	60 ft.
S gauge	82 ft.
O gauge	110 ft.
G gauge (1/32)	165 ft.
G gauge (1/24)	220 ft.

Don't expect extreme precision performance from an economy line product. Some hobby shops avoid selling certain lines of trains because they don't want their customers to get stuck with locos that don't perform well. A well made locomotive should run smoothly and be able to get down to a crawl without jerking. Reliable hobby dealers will stand behind the equipment they sell.

Locos with two wheel pickup are more prone to operating problems than multi wheeled locos picking up power from six or eight wheels. Hobby shops sell kits that provide multi-wheel pickup for improving operation. Nickel silver rail tends to stay clean longer and is easier to clean than some other types of rail. Locos should be lubricated periodically but avoid over oiling.

Television Camera Locos

Now available in scales from HO to G are locomotives equipped with miniature television cameras which show the right of way as the train moves forward. TV locos finally put the engineer into the cab and let him see the right-of-way as the train proceeds on its route. Be careful that your tunnels are fully scenicked as hollow-shell tunnels show up as plain as day. Televsion camera locos operate best in well illuminated train rooms.

Automatic Couplers

Automatic couplers are a must if you do switching. Happily most ready to run equipment come with automatic couplers. Many advanced workers in all scales prefer to use Kadee couplers available from your hobby shop. Kadee couplers install easily and are made with a variety of coupler shanks to fit almost all makes of equipment including European equipment made with the NEM-362 swallow tail shank. Kadee couplers look almost like the real thing and are available in a wide variety of scales. Tinplaters (Lionel, K Line, American Flyer) usually stick with couplers compatable with Lionel or American Flyer You'll want to use the same coupler your friends use so you can interchange equipment.

From Plan to Layout

Photo by David L. Waddington

by the layout doctor

A track plan is basically a series of turnouts, curves and tangents, but getting them to all line up takes a bit of know-how.

A reader wrote to ask how to transfer a layout drawing to actuality. I started to answer but realized that any brief answer would be incomplete and not too satisfactory. I gave him a few ideas and told him I'd have to write a whole article on the subject; here it is. I am assuming for simplicity that you are going to work with a table-top layout, a platform, whether Homasote or plywood. The same ideas, however, can be adapted for use with an open top benchwork, or to cutting roadbed from Homasote which you have to mark out first with a pencil.

Nearly all layout plans published use the single line to designate both rails of one track. This represents the centerline of the track which is imaginary, though sometimes indicated by nail holes left in plastic ties. All layout plans printed are to some scale, usually indicated by one foot square grid lines superimposed on the trackplan. We also try to show the radii of the curves, supplemented by information in the descriptive article. We also try to show the major curve centers.

The first part of any layout to be put on the table is any or all groups of switches in the main lines. On my designs, I try to keep switches all at the front, or within reach of access holes. You do not fasten them down, but you do connect them with rail joiners so they can be moved around as a unit within a small area. Do not wire the switch motors until later.

The second thing to be done on a model railroad is to layout the main curves. These will later be connected with "tangents" or straight track.

There are several ways of laying out curves. If you are using plain wooden roadbed, the curves are already made and are designated as being of a certain radius, so you don't have to worry about curve centers much.

Special templates for popular curve radii (as well as straight) designed to fit between HO rails and keep the right curvature used to be available. I would guess, however, that more than half the model rails in HO use neither of these aids: they use cork, rubber, or *no* roadbed, or even glue down individual ties themselves. For these fellows, drawing the center line the table with a pencil is necessary.

The tool of most help in making these "big" curves is the trammel, Fig. 2. This is a home-made one, formerly a yardstick. Drill a hole near one end, which fits over a nail located at the curve center, and then drill additional holes 18", 20", 22", etc., out and you are ready to draw curves. When you nestle half the rubber or cork roadbed against the pencil line you have determined the curve. You can also hold the center of the track or individual ties to it.

In Fig. 1, I show part of a fair sized table top layout such as I like to draw. I seldom show the distance from each edge of the curve centers as I do here, but I usually do indicate the centers and the radii of the curves.

After your first step in making such a layout (1), grouping together the important switches, you can swing the 20" radius curve, (2). Its center is right on the table and so there should be no complications except reversing the trammel so that excess length does not interfere with the wall.

Curve (3) of 24" radius has its center in the open space in the center of the layout. The idea shown here is to tack a piece of 1X2 across so that the center will fall on it and you can place your nail. The 14" radius curve at lower left also has its center in the open space, but in this case you might try using a stool (plus books or a box to table height) to locate the center. Curve (5) has its center not only off the layout table but somewhere out behind the walls of the room. It is a special problem which I'll explain later, though you would have to make the curve at this time.

Curve (6) can be swung at this time, using the stool method unless the center happens to come on the front part of the model railroad table, in which case you could use the idea in (2).

To fit in a level crossing as at (7), you first have to measure some angles. Extend the tangent at left and measure the angle (7A). On the drawing I get 20° . A protractor should be sufficient for this measurement. If you intend to use a 45° crossing at (7), that leaves 25° to be made up with the extension of a tangent at (7B), although the curves at the other end of that tangent, (8) and (9), have yet to be laid out.

The curve centers of (8) or (9), which are now of different radii, are indicated by dimensions from wall and table edge. Actually, however, one of them would have to have its exact center found by hunting around. Move the stool around until you

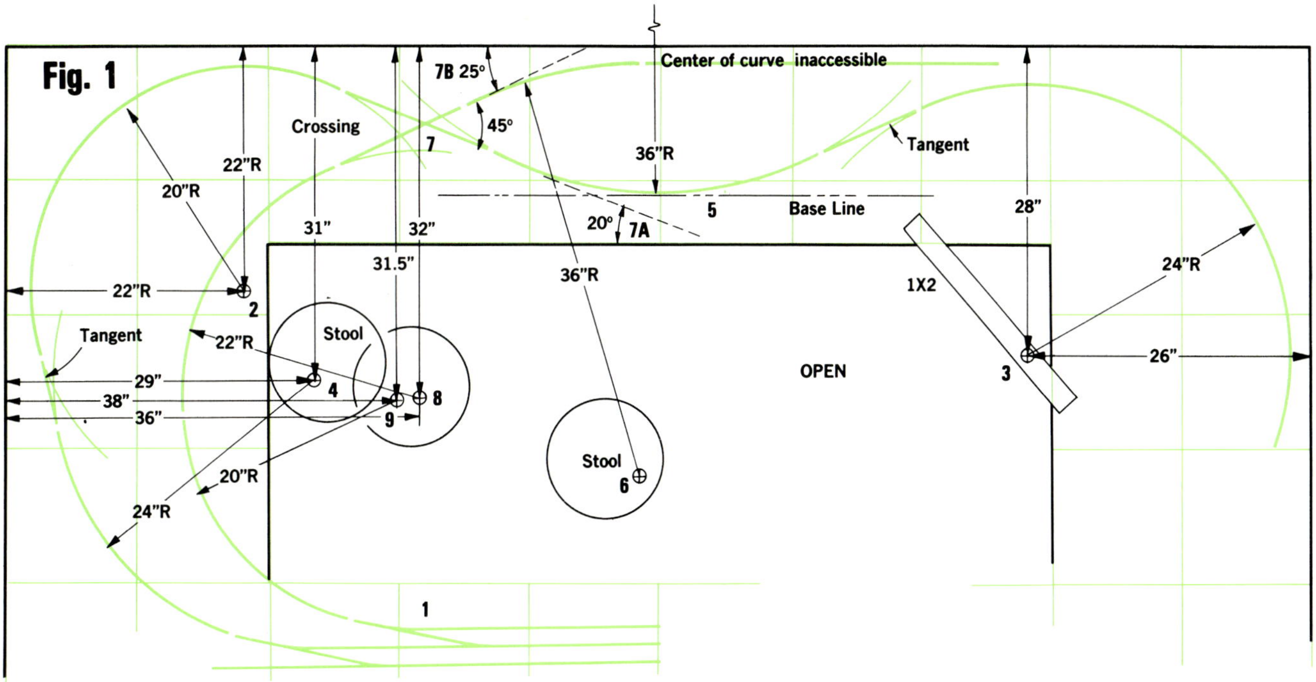

Photo at left shows the Bay State Society of Model Engineers' new O scale layout under construction at Boston. Though more complex than layout discussed in text, basic idea is the same: for a pike that operates smoothly, plan ahead!

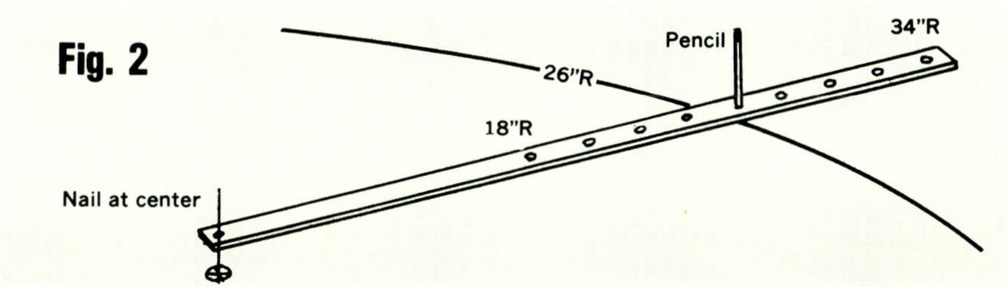

can swing the curve to line up smoothly with trackage at each end.

Let's get back to curve (5) with its center not accessible. This can be laid out best with offsets which is what a surveyor would do. They have real railroad track design books which show tables of offsets for different degree curves, though not for radii in inches as modelers use. You can figure out your own offsets by graphing the curve as you used to do in algebra. Turn things around so that the unavailable center of the curve is at the "origin", where **X** and **Y** axes cross. Let the edge of the table and the base line for offsets be parallel to the **X** axis.

Since the equation for a circle with its center at the origin is $x^2 + y^2 = r^2$, you can substitute the real **r** in that and continue as in Fig. 3B. Then you make up a little table of x and y values as in 3C. No need for great accuracy: I took it every 4" for x. First you get the y value as a square root. I show these values indicated on the graph. They must then be subtracted from 36 to get the offsets at each 4" increase in x. Just make your base line and at 4" intervals measure back 0.2". 0.9", 2.1", 3.7", etc., as indicated. For the other side of the y axis, these would be the same (see 3A).

Connect up the points so determined with a curved line. Here is where a loose yard length of rail comes in handy since it curves nicely.

Though there are other points of information which the beginner will find useful when he constructs his first layout, the above procedures are sufficient to get the project off the drawing board and into the "hardware" stage. One last hint: take your time and work accurately. Don't "cheat" to force a curve to fit, since this will likely result in a kink in your finished trackwork and likely "mysterious" derailments at that point in the future.

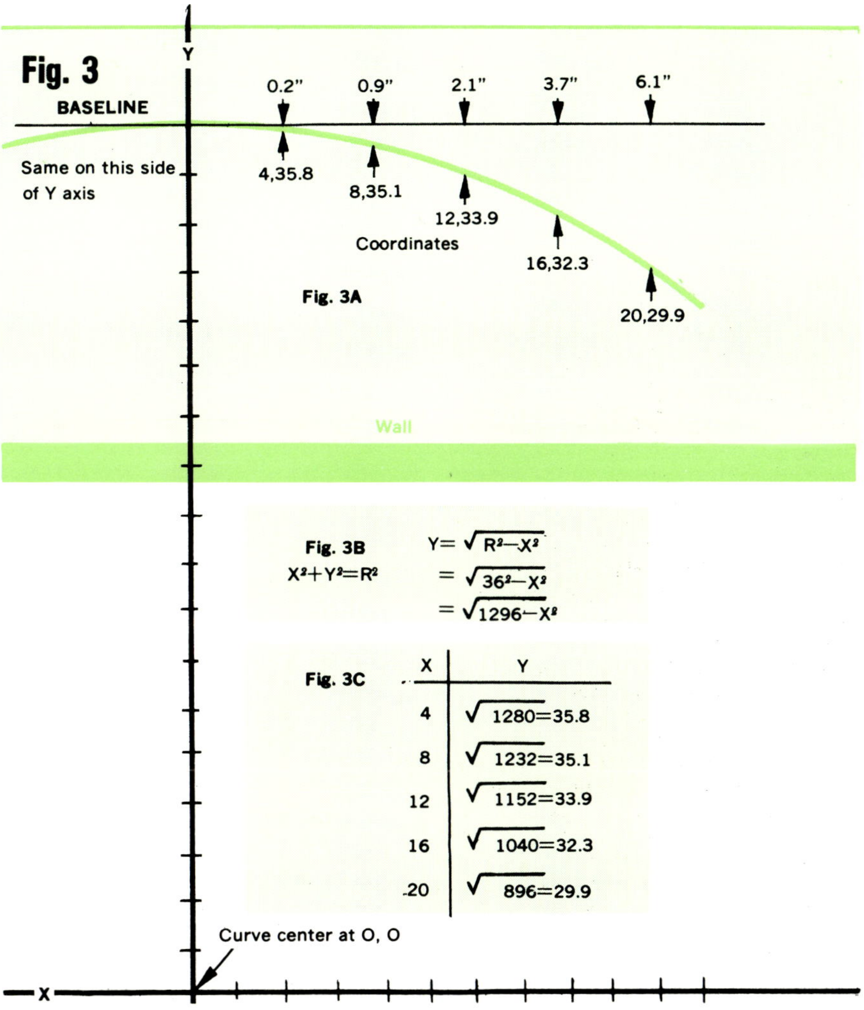

Fig. 3B

$$X^2 + Y^2 = R^2$$

$$Y = \sqrt{R^2 - X^2} = \sqrt{36^2 - X^2} = \sqrt{1296 - X^2}$$

Fig. 3C

X	Y
4	$\sqrt{1280} = 35.8$
8	$\sqrt{1232} = 35.1$
12	$\sqrt{1152} = 33.9$
16	$\sqrt{1040} = 32.3$
20	$\sqrt{896} = 29.9$

Here's a time tested and basically sound layout design providing the basis for an interesting yard and terminal complex - an discussion on track plan design.

by the layout doctor

Given a large area, a small gauge—in this instance, HO—some model rails become quite unrestrained in designing a pike to fit this huge space. This month I present just part of such a design recently submitted to me: a reverse loop and terminal complex in an area of 9' x 11' plus connecting tables—think of it as being about the size of your 9' x 12' rug. It is a prime example of an over-designed layout, shown in drawing marked ORIGINAL TRACKPLAN DESIGN. I have partially modified the plan to be more within the rules for good layout design, yet keeping the same general idea in mind; see REVISED TRACKPLAN.

To begin with, though this will not be apparent from the drawing reproduced here, the plan was drawn—and very well drawn, too—to the odd scale of 1-1/4" to the actual foot. Why? Obviously, to fit it on the available paper and drawing board. And, in partially retracing the plan, the new design had to be to the original scale, again not apparent to the reader since plans are always reduced to fit RMC's page size.

When *you* draw a layout for your own use, try to use a more suitable fractional scale. A very popular scale is 1-1/2" to the foot, making each 1/8" on your ruler represent 1" on the layout itself. For a pike the size of the terminal shown here, this gets pretty unwieldy; my thought is that the plan of any model railroad should be readable without having to fold or roll it to keep it "handle-able." If 1-1/2" scale proves too big for your paper or drawing board, give some consideration to 3/4" scale, where each 1/16" represent 1 actual inch. In between, there is another "offbeat" scale which, because of its frequent and constant use, has become quite acceptable. This is 1" scale. And to get 2", as in track spacing for HO, you can estimate by eye what is 1/3 of 1/2" and come close enough.

Refer to the main drawings. The *original plan,* first of all, requires that switching in the Inner Yard, as well as the Passenger Terminal area, be done using as a yard lead the single approach track into the tunnel. Now this is poison, even though in prototype practice the yard at the Swiss end of the Gotthard tunnel has a switching lead running into an auxiliary tunnel alongside the main bore. Trains derail and uncouple enough in switching without having them do so hidden from the human eye and hand. Here, a wreck in that yard-lead/tunnel would tie up the approach to the main terminals. You'll note that on the revised plan I have made a double approach to the neck of the big loop around the terminal area. Thanks to the double-slip (more about *that* later), either of these tracks can be used for a yard lead while the other handles regular traffic.

In addition, on the revised plan, I have changed levels somewhat to have the line to C cross under, rather than over, the yard approaches, making them more accessible. Indeed, having both out in the open instead of in a tunnel will prove much handier for switching purposes where the hand must often lend an assist—to jog a train, or to rerail or recouple a car. If you must have the line to C cross above the two tracks, go ahead. You could also (by moving it off the curve) have it cross the approach tracks at grade on approximately 25° or 30° crossings.

On the original plan, I have had to do some hard studying and guesswork to figure out just what certain tracks are for. The labels therefore are mine. The engine ready tracks would be to store engines out in the open that are to go out on another run soon, in contrast to a roundhouse where engines are stored live or dead, for longer periods. An engine ready track area used to be a great sight for photographers, what with all the smoke and steam—and still is, but with diesel fumes instead.

But the way the engine ready tracks were originally drawn meant using too many switches: you have to have two switches, one at each end of such a track, which will at best store only two steamers or one multi-unit diesel. Possibly the back switches (to the left) could be eliminated at some savings. However, on the revised plan I have made the engine ready tracks the open radial tracks around the turntable. This might not be acceptable on a real railroad, to be sure, since an out-of-order turntable would tie up all locomotives.

The original turntable has its radial tracks into the roundhouse marked as being 9° apart. This would be O.K. if you are sure you would design and build your own roundhouse. But, if you have any idea that you might use a Vollmer roundhouse, bear in mind that it is made for tracks 15° apart, as on the Fleischmann turntable. (The AHM turntable has tracks 20° apart to fit Rivarossi roundhouse sections.)

By eliminating the engine waiting tracks, I have, on the revised plan, been able to move the throat of the Inner Yard nearer the upper control position. One very bad feature of the original plan which crop up again and again is that switches would be out of reach, even from additional small

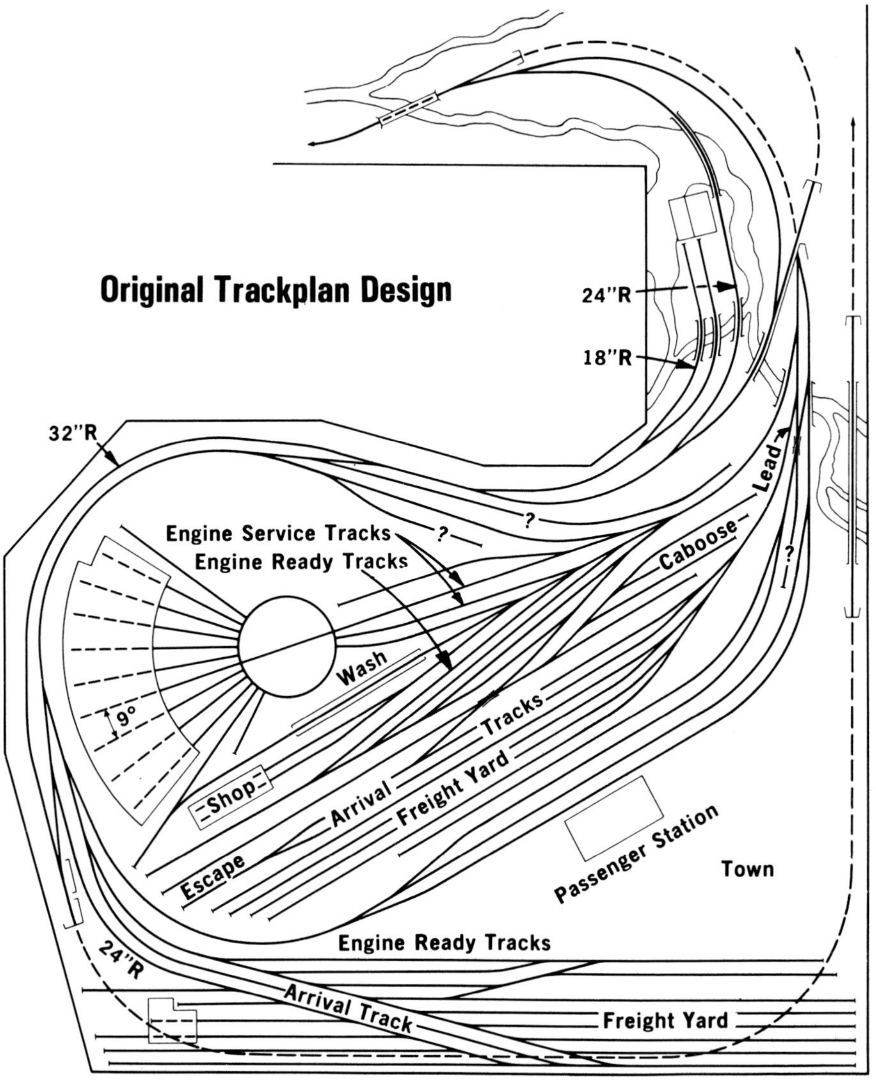

The Overdesigned Layout

"pop-up" control holes. "Out of reach" for switches would be around two feet back from the table edge, regardless of gauge, and even that is quite a reach if there are poles, signals, etc., poking you in the abdomen.

Similarly, I have elimated what I took to be caboose tracks, and replaced them with just one such track. Additional cabin cars could be stored on the body tracks of the yard.

I have replaced the two arrival tracks of the original plan with just one, with a similar escape track over to the engine terminal area. But, as I study my revised plan, I suspect that I should eliminate the escape track, since it involves two switches almost three feet in from the edge. Let the road engines uncouple and pull up to a dead section of track, while a switcher moves the cars to another track.

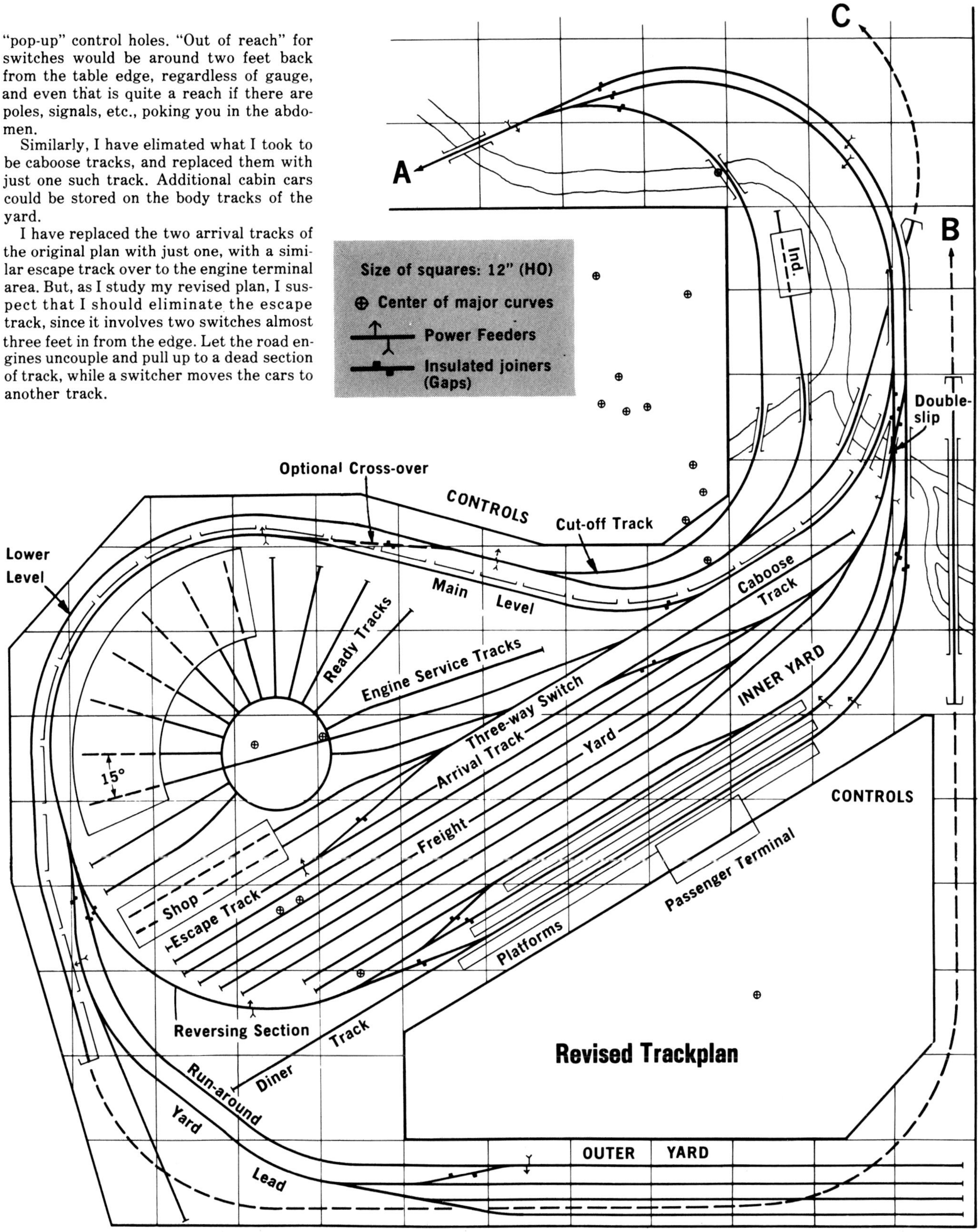

Revised Trackplan

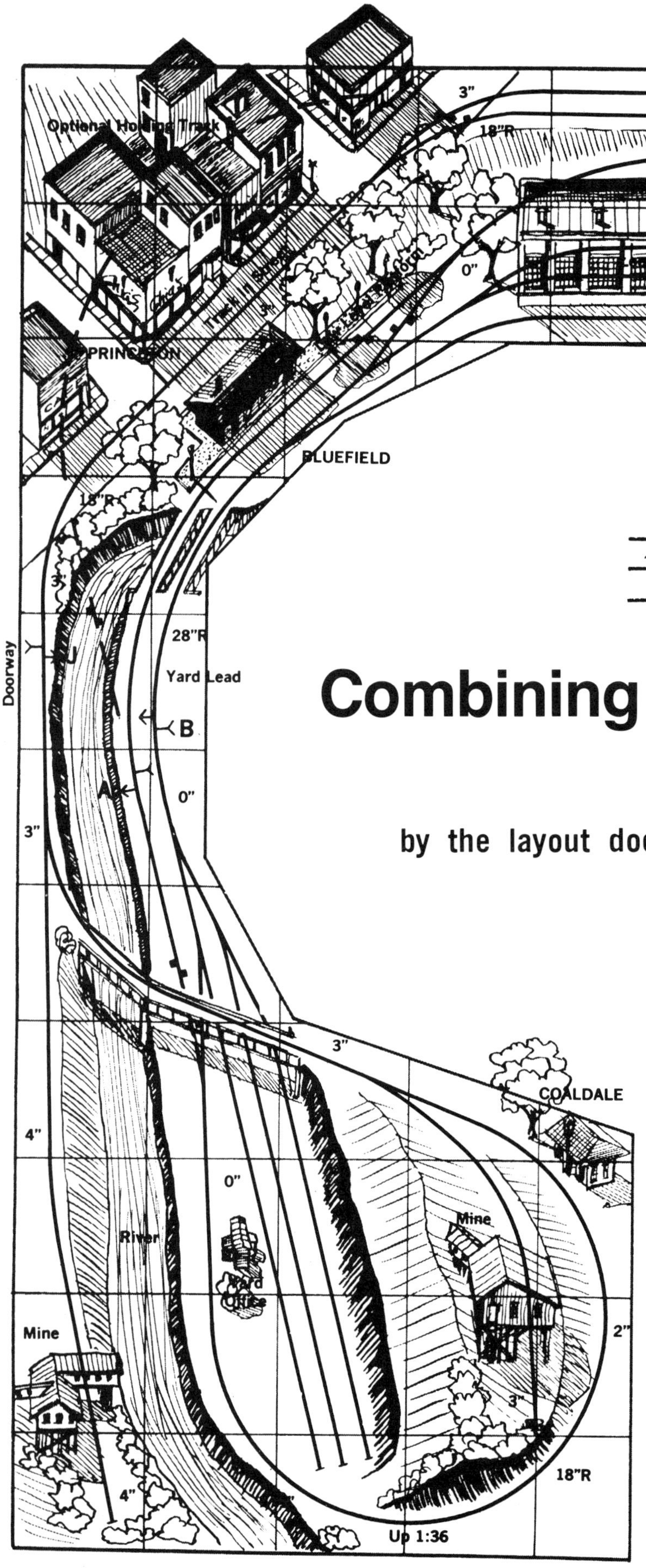

Combining Parts of Layouts

by the layout doctor

Layout Symbols

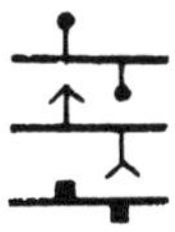

WESTERN VIRGINIA RAILROAD

NEW JERSEY TRANSIT

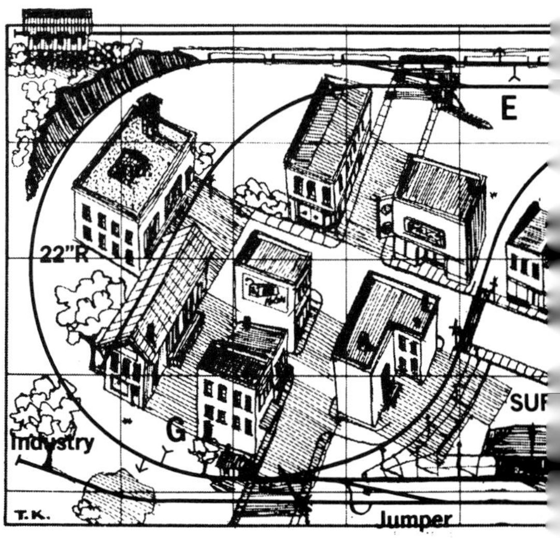

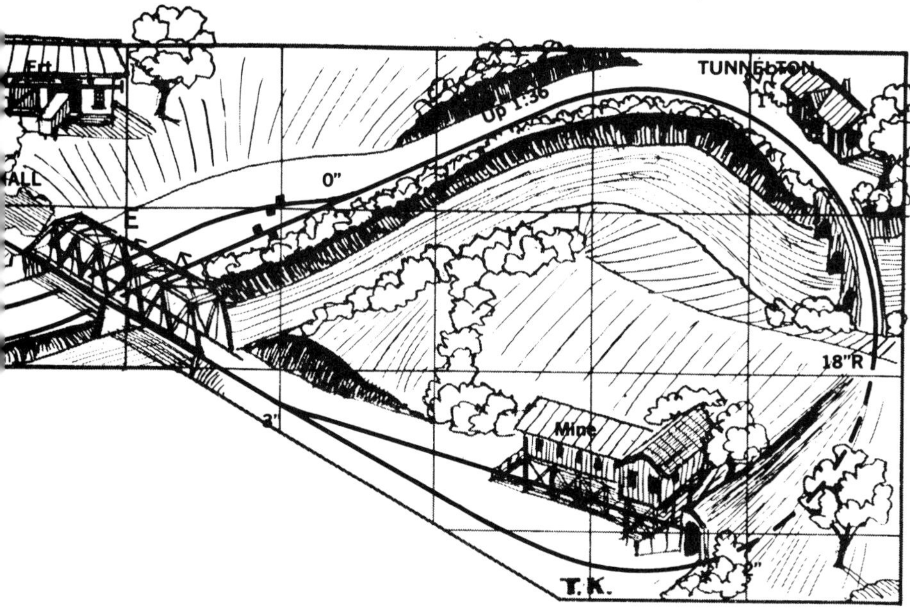

One track plan may not contain all the elements of your dream railroad but combining parts of two or more layout plans may help you come closer to the layout you want.

One big help in planning your own layout is to keep a file of all the model railroad magazines you can acquire so that you can pour through them, perhaps with the help of an NMRA Periodical Index, for ideas on whatever subject you may have on your mind (some tear out pages and articles to form folders of unmounted plans and articles on track layouts, for instance). Of course, it is advisable to keep any books or leaflets on layouts, as well as pertinent pages from any overseas model magazines you may acquire. British layouts are generally quite good and applicable, Continental ones less so.

Plans which have portions which appeal to you should be marked somehow, or gathered in a separate folder, perhaps with a few notes or doodles by you. When you are ready to plan your own layout, you can review these special ideas to see how they can be combined. I'll take some specific examples to show you how it can be done.

In the July 1961 RMC I had an article on straight-line point to point layouts, the most sophisticated of which was for the HO Princeton-Bluefield Interurban. This was a narrow sixteen foot long trolley pike which used overhead wire in the city street out to the first passing track, as well as in the car barn area at the other end of the line. All of the rest of the line was intended to be run from outside third rail, *a la* Philadelphia & Western, Scioto Valley, Puget Sound Electric, etc. I've always liked that idea and swore that if I ever built a fourth trolley pike I would use it, since automatic changeover devices can be devised.

I had a page drawing of several ideas utilizing the insides of what I loosely called reverse loops to add scenic interest to any layout or extension. One of these, "Coaldale", I combined with part of "Walnut" to finish off the Princeton-Bluefield pike into a complete circuit so that continuous running could be done. Since I know everyone is not an interurban fan, I enlarged minimum curves for HO to 18"R with some 30" and 28"R for variety. Thus, steam or diesel locos could be operated, or wire and/or third rail could be installed similar to the original Princeton-Bluefield.

The result is an 11' x 15' along the walls, L-shaped pike which should give a lot of fun whether built for traction, steam, diesel, even narrow gauge—or any combination thereof. What might represent 20 or 30 miles of railroading is compressed into a run of slightly less than a one scale mile as is always necessary in the geography of model railroading. Suppose we call it, for reference, the Western Virginia R.R., with reporting marks WV and a herald of W over V so located that the diverging arms of the V become the outer arms of the W.

Since the lowest tracks of any layout should always be built first if they are to be covered or obscured, and since the most complex switchwork should be built first, it works out nicely to initially construct that part of the WV that is at 0" level—i.e., blocks *A* and *B*, including the yard lead, yard, and engine terminal from the yard office at lower left out to Barnes and just beyond. If you think you should have the optional holding track under Princeton, this is the time to install it. One more 4R and one more 4L switch will be required, but I believe I have tabulated enough flexible track to build it. It would add yet another control section, not marked, *K*.

Since a steam/diesel railroad is likely to need to turn locos, whereas an interurban might not, I have included a turntable which could be left out if no steamers nor diesel A-units were to be run. I have also reversed the "car barn" in changing it to a loco shed, so that it is fed from the Bluefield yard lead, section *B*.

Proceeding to the right, toward Tunnelton (note that I have retained station names from the "inspiring" layouts), the big change is up at Marshall, where the passing track has had to be made longer. Then,

See bottom Page 17 for map of Bluefield-Princeton Interurban Railway

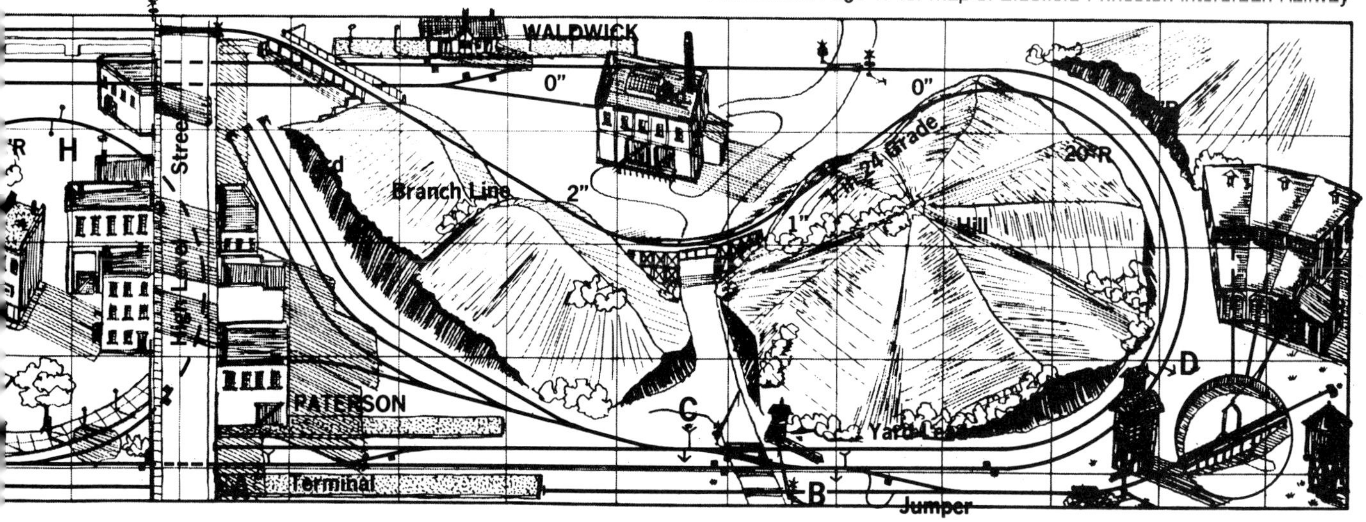

LAYOUT PLANNING IDEAS

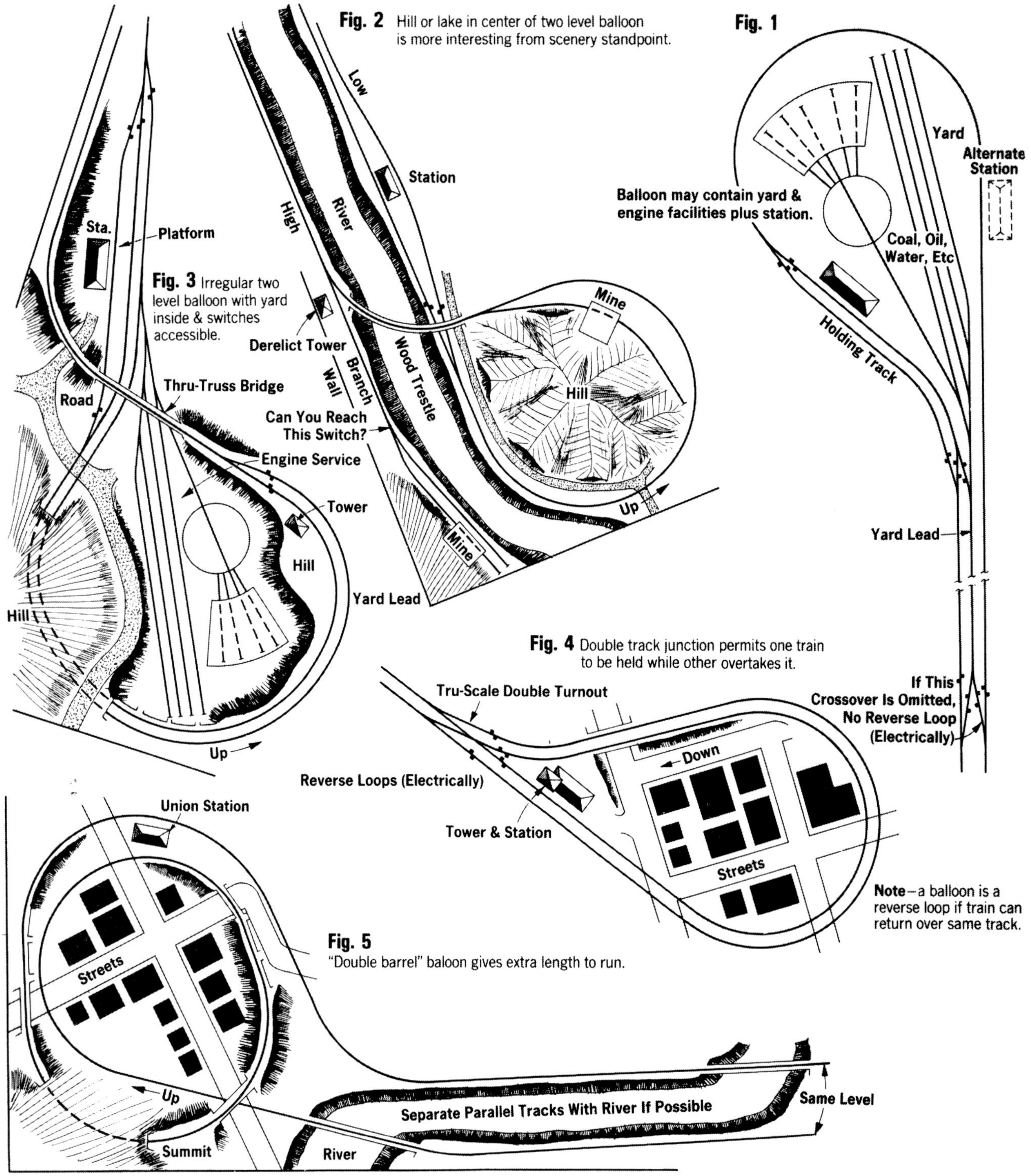

Fig. 1

Fig. 2 Hill or lake in center of two level balloon is more interesting from scenery standpoint.

Fig. 3 Irregular two level balloon with yard inside & switches accessible.

Fig. 4 Double track junction permits one train to be held while other overtakes it.

Note—a balloon is a reverse loop if train can return over same track.

Fig. 5 "Double barrel" baloon gives extra length to run.

passing down the main street at Princeton, the tight wye for turning interurbans has gone, though you might like to show it as abandoned trackage. There are plenty of examples of a main railroad track passing down the middle of Main Street, but of course you shouldn't whip your locomotives through at high speed. The station sits off to the side and also serves as a high level station for Bluefield.

The trackage swings across a creek and some lower tracks to Coaldale, then down a hill to enter the Bluefield yard from the other end. On this part of the pike there are two mines on long sidings which could be coal, ore, or some of the Sweetnin' E.L. Moore finds down there in Carolina. There is a third mine at the far right extreme of the layout, and the freight house siding at Marshall could be extended to serve yet another.

I tabulate track requirements as follows for HO: 8 RH #4 switches; 7LH #4 switches; 1 #6 RH, 1 Y-switch; 99' of flexible track. While switch requirements would not change for N gauge, you'd need from one-half to three-quarters as much track, depending on the reduction in size of the table; in O gauge, you'd need twice as much flexible track for a double sized layout, of course.

For control purposes, I'd suggest the system developed by Atlas (or its equivalent) using Selectors, of which you'd need three, and two power packs. You would then be able to control two trains independently, which probably would be all you could handle, even though with three passing tracks you should, in theory, be able to run three trains at once. It is unlikely that such a small pike would have, even during its halcyon years, more than two engines in steam at one time.

The grades on the WV are all 1:36. That works out to 1/3" rise in 12", which is close enough to a 3% grade. Pretty stiff, yes, but within the capabilities of nearly all HO engines if trains are kept short—and short trains are almost a must on a relatively small pike such as the WV to make the pike itself appear longer.

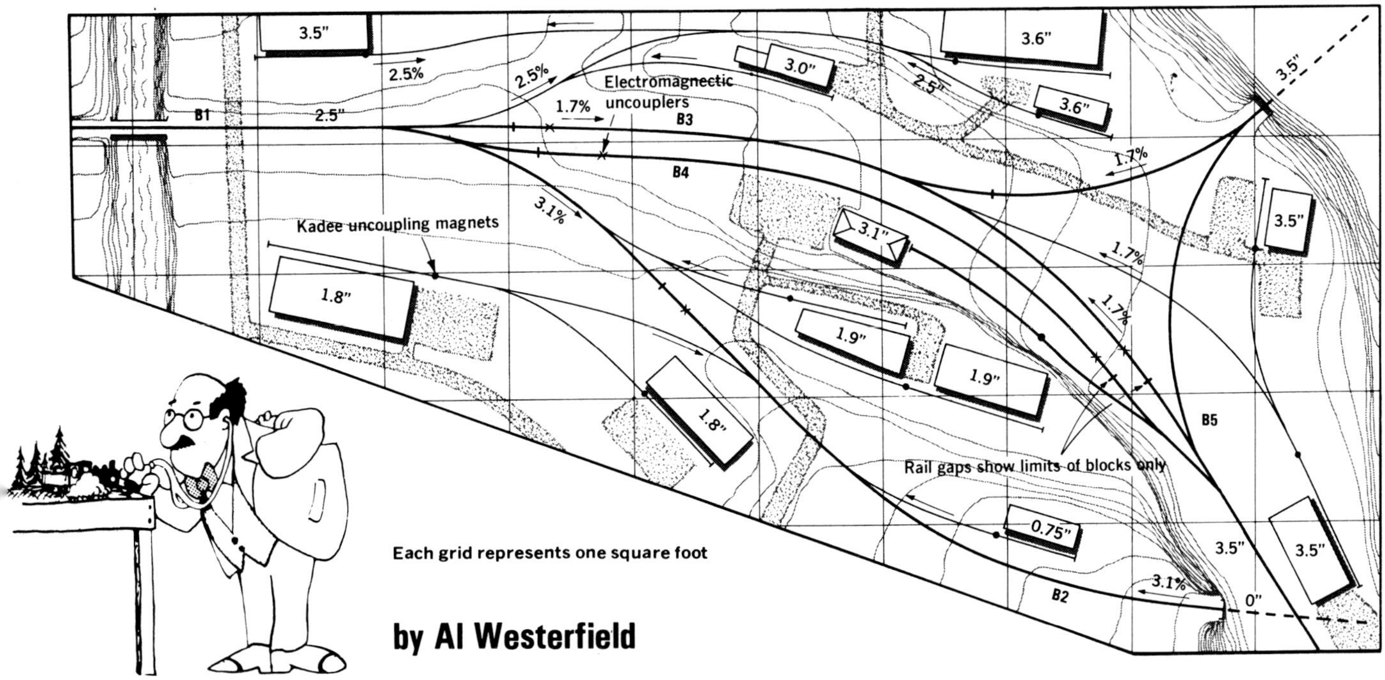

Each grid represents one square foot

by Al Westerfield

A Corner Multi-Level Switching Module

A multi-level switching module avoids the cluttered look of most switching pikes and can later be incorporated into a larger layout.

A friend asked me to come up with some ideas for an odd-shaped space he had set aside for an HO layout. I got a little too involved and wound up designing the complete layout for him. His area was 10'-6" long by 2'-0" on one end and 5'-0" on the other. The only requirements were that it be a switching layout with plenty of action, but without complicated trackwork. Knowing the modeler helps to fill out the picture, however. I knew that he lays his own track, including turnouts, so force-fitting the plan to standard curves and turnouts wasn't necessary. Scenery is one of his passions, so the plan had to include plenty. Finally, he runs nothing larger than Consolidations, so turnouts as small as No. 4 could be used.

Then I added my own prejudices. Mainly, I hate switching layouts. Therefore this would hopefully not *look* like one even if it was. Most switching pikes I've seen contain lots of straight track running back and forth on a flat board. That's fine for switching contests at conventions but hardly holds attention for prolonged periods of time; the layout is completed in a week or two without challenge or a chance for real modeling skill.

This layout would depart from all the above cliches. First, almost all trackage would be on grades with different elevations for different industries. Second, very little of the track would be straight. The combination of these effects fools the eye into believing that the track is longer and the industries more separated than they really are. Finally, there must be a logical reason for a concentration of industry. Bitter Creek Junction is the result.

Bitter Creek is the headquarters for a medium size connecting railroad. The main line, running east-west, is joined by northern and southern branches at this small manufacturing town, providing a natural area for industrial facilities. The northern branch joins the main via a wye. This artifice makes the most of the corner area of the room while providing a turnaround for motive power. The main has a passing track to facilitate switching moves. All dead-end main and branch tracks have room for at least an engine and car; some have considerably more.

In HO each square on the plan is 12"x12". All grade percentages are given by an arrow (indicating where the grade begins) with a number near it. Elevations for all industries and level portion of track are shown. X's represent placement of electromagnetic uncouplers, and dots represent permanent magnets. Five blocks are shown, as dual cab operation is contemplated.

A dozen industries are provided. Notice that their elevation increases with increased distance from the viewing area. The surface area therefore is considerably increased. All industrial trackage where cars will be spotted is level. Building placement has been roughed in to show the approximate sizes possible for each location. Some can be over 18" if desired. Since the natural flare of trackage is to the right of the plan, most sidings are entered eastbound. Enough switch-backs and westbound trackage is included to keep it interesting, however. All turnouts are approximately No. 4.

Uncoupling on the southern branch grade is accomplished by a movable wire between the rails. When in the up position it catches the axle of the car, providing the necessary slack for the uncoupling ramp. It can be actuated by a switch machine or choke cable. The grade on the main line passing tracks is slight enough to make this kind of installation unnecessary.

Most areas of the layout are reachable from the front. By making the northern branch mountain a facade, the remainder becomes accessible. The topology has been approximated in the plan. Obviously, by varying the pitch all sorts of scenery becomes possible. Cliffs, gentle slopes, retaining walls and a small stream are indicated as potential scenic interests.

Electrically, the layout is designed for ease of operation, yet permits a second operator to work simultaneously. Two sides of the wye are wired to Block 5 to reduce polarity problems. The wye area could keep one operator busy all night. Consequently, most industries are controlled by Blocks 1 and 2. They can be worked without going onto the passing sidings at all. I will admit to throwing a slight curve ball into the simple track work—the crossing at the east side of the wye. It does permit more complete space utilization, however, and crossings are not that hard to build or wire.

The overall result is a layout which requires a modicum of skill to build and a good deal of time to complete. With trains approaching from four directions to service 12 distinct industries, the operational problems can be as complex as desired. It incorporates the advantages of a larger layout into a small space without being overly crowded. In addition, with minor modifications it can be used later as a corner of a larger layout. Since no exotic trackwork is involved, with a little fudging commercial turnouts could be substituted. But, best of all, it doesn't look like a switching layout.

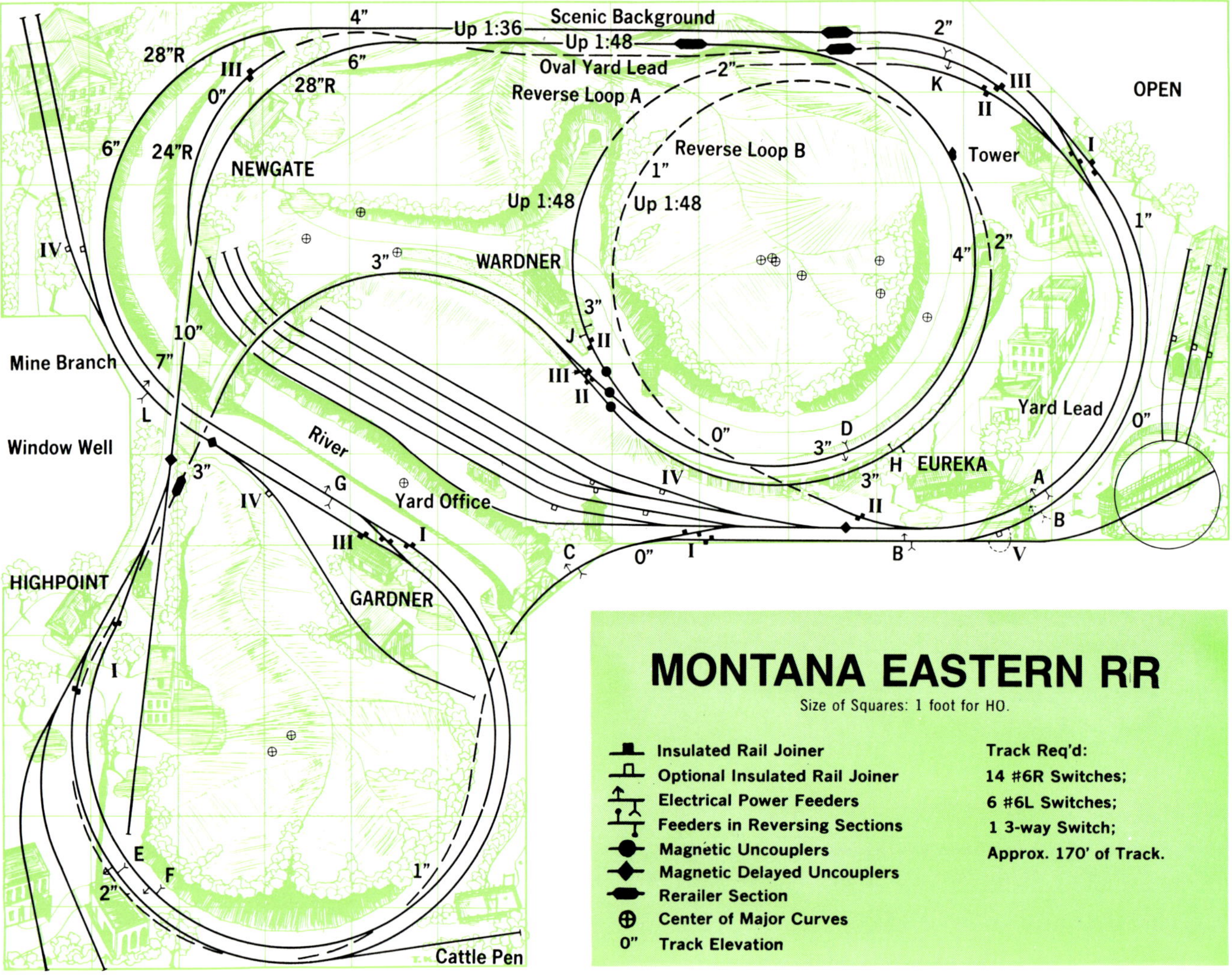

Insulated Rail Joiners

A scenically- and operationally-interesting track plan is used to illustrate the various "rules" about insulated joints.

by the layout doctor

It is part of the nature of 2-rail model railroading that from time to time you need insulated rail joints, for reasons which will soon become apparent. In fact, they are useful even on 3-rail layouts to isolate one running rail for signaling or working a crossing gate—remember the fibre pins and matchsticks?

Insulated rail joiners for HO rail were first made by a firm called E&R in Colorado. They were similar to those now in use: transparent plastic but with a separate red fibre "wall" riding in the joiner to keep the rail ends from butting each other. This was in the very late forties. Soon Atlas improved on this design by molding the insulating wall into the joiner so it couldn't fall out and get lost. The insulated joiners haven't been changed much since, except I note that some are molded in colored opaque plastic, which has both good and bad points.

What did railroaders do before the days of insulated rail joiners? (1) They simply left "gaps"—a term you still hear—in the rail while laying it, about a 1/32" break in the continuity of the rail. Rarely such gaps would get closed by expansion, by dirt, or by construction elsewhere but generally they did quite well if the rails were held down solidly by four spikes each side of the gap, and the rail, if curved, was pre-shaped not to fight out of position (both good ideas with insulated joiners, too). (2) They would build the whole layout with no gaps at all, then cut in the gaps (after their locations were determined) with a cut-off disc.

Hard motor tools (such as those made by Dremel) make the task of cutting "gaps" quite easy. Before cutting gaps with a saw, put at least four spikes on each side of the proposed gap, and even then some adjusting will be required to keep the cut rail-ends in line. In real tough instances, use a brad instead of a spike on the outside, and solder the rail to it, taking care not to foul up the plastic ties, if any.

The need for insulated joiners, gaps, or what have you, is shown by the attached drawing of the Montana Eastern R.R., an HO pike 14'-0"x10'-6".

While most of the main line switches are accessible from the front of the pike, there are two pop-up holes: a triangular one in the upper right corner to reach switches and scenery there; and a rectangular one at extreme left for the same reason (as well as, originally, to have access to a window there). The window access spot could even be equipped with auxiliary controls to run the little branch serving two mines (electrical section L).

In putting in insulated joiner locations on this layout, the following "rules" were used: "Rule I" notes that any track or block that sometimes had to be under separate control had to be separated from adjacent blocks by insulated joiners. Thus, the looped joiners at each end of the Eureka passing track are accounted for, since obviously when trains pass, they will run in opposite directions. The same rule applies to the Gardner-Highpoint passing track at lower left. There are 4 control sections; between these are two stretches of single track: Highpoint-Newgate-Wardner-Eureka, and Gardner-Tower V-Eureka.

Any track that is a reversing section must be isolated from other tracks by insulated rail joiners; thus Reserve Loop A (Wardner-Tower V) and Reverse Loop B (Wardner-Eureka) must be separated by "Rule II," so marked.

"Rule III" calls for the separation of any track which *can* be separately controlled for more variety in operation. Thus, the short tangent between the two switches at Wardner has insulated joiners, since once a train has been diverted to either reverse loop at Wardner, there is no reason why another train cannot follow or oppose it on the main track. Under Rule III, the mine branch cutting off at Gardner is separately controlled, isolated by insulated joiners. It *could* be controlled by the inner passing track (F) at a slight loss of flexibility, scarcely noticeable for one-man operation. The back part of the oval yard lead (G) is similarly separately controllable under Rule III.

The back of the oval yard lead which wanders around under the upper tracks partly for clearance, partly for better visibility on the plan, *need not* be separately controlled. On the Montana Eastern plan, insulated frog switches (Custom-Line or Casadio) have been specified. If they are used, (G) will not be needed (nor insulated joiners at each end thereof, though they might prove handy to have). The insulated joiners would also be handy to have using all-rail frog switches. However, there *must* be at least one pair of insulated joiners in this oval even if there is not a separate control section (G); otherwise, throwing one of the yard switches or one at Tower V would short out the whole yard lead oval. The pair at left, below Newgate, might be retained or move them to a location midway along the back.

The numerous open squares on one rail only are *optional* insulated joiners. This is because insulated frog switches do not "kill" the track they are set against, and engines will run, car lights will light and mysterious short circuits may occur from hidden derailments. By putting an insulated joiner in one rail (and in the corresponding rail in each parallel track), toggle or rotary electrical switches may be used to energize or de-energize (kill) such yard tracks. This isn't always worth the trouble, unless an engine is to be held on that track while another engine works elsewhere in the block. In a parallel track enginehouse as at extreme right, each engine space should be killable so that an engine may be operated from the front space right out to the turntable without a second engine stored on the same track following right along. Hence, insulated joiners are needed in enginehouse tracks even with all-rail frog switches, which in a yard situation kill tracks they are set against and feed power only to the track they are set for. The optional rail joiners sometimes necessitated are "Rule IV."

There is one more rule which did not crop up on the ME, applying only to blocks containing all-rail frog switches. Modelers have gotten quite free and easy with the wide use of insulated frog switches, of putting the feeder wires to a block almost anywhere. With all-rail switches, however, all switches must point *toward* the feeders, as on the Montana Eastern. However, sometimes this is hard to do without a lot of trouble, so a feeder is placed behind a switch. Temporarily relocating feeder (B) past the roundhouse switch, which is shown dotted to ease confusion, illustrates this.

If the roundhouse switch were then thrown straight, (B)-dotted would be directly shorted out. Accordingly, an optional insulated joiner is shown with a V (Rule V) beside it, and with a dotted jumper wire around it. A block can be fed from behind a switch with all-rail frog switches by observing the precautions noted.

Notice that, excepting under Rules IV and V, insulated joiners usually come in pairs, one in each rail. Since flexible track and commercial switches are so widely used, these insulated joints come approximately opposite each other, though there is usually no need for them to be exactly opposite. To make the drawing a little clearer, they are drawn offset at crossovers, such as Tower V.

There is no need to have insulated rail joiners at switches except to follow one of the rules given here. Between switches of a crossover, Rule I applies. Near the switches to and from a reversing section, use Rule II. Rule III usually works out to place the insulated joiners near switches. It is also true with rule IV, for why waste yard body track length, or why cut a rail a few inches long in order to move the insulated joiner away from the yard switch? Rule V is the only situation where you *have* to have one, and it really could be farther from the switch than it is. And, as insulated frog switches come more and more into use, the need for the Rule V situation diminishes.

How are all of these separate blocks connected to the power pack? That is another story which I'll have to go over at length. Briefly, one can simply put a separate power pack, or separate set of speed and direction controls, on each block (A) through (L), but it is often more economical to use one or two packs and wire up some kind of cab control. One form of cab control, suitable for the independent running of two trains, is that expounded in the Atlas Tool Co. book on wiring which your dealer probably has. Of course, this system utilizes Atlas components which make up into a very compact control panel.

The Montana Eastern has several interesting scenic vistas. A person inside the L, say at the main control, can look toward the window and see four separate levels of track, plus the river: the yard and lead near 0", the curving bridge across the yard at 3", the apparent double track to the left of Gardner at 7", and the high trestle from Highpoint to Newgate at 10". A person standing inside the angle of the L and looking along the river can see the above, and also the mines in the upper left corner. Standing inside the left arm of the L he can see, behind the mountain, the mines at Highpoint at lower left.

One of the less attractive views would be from inside the mountains along the front of the scenic backdrop. This would have to be left open somehow, even if not to daylight, for access to the back tracks as well as to the two reverse loops. The inside of the lower left mountain would also have to be open to reach the hidden track Eureka-Wardner. If you did not want or could not have the window access space, the left end of this under-mountain opening could provide access to the Highpoint switches. In this case, the #6R switch at the mine at upper left would have to be left out since it couldn't be reached

BLUEFIELD-PRINCETON INTERURBAN

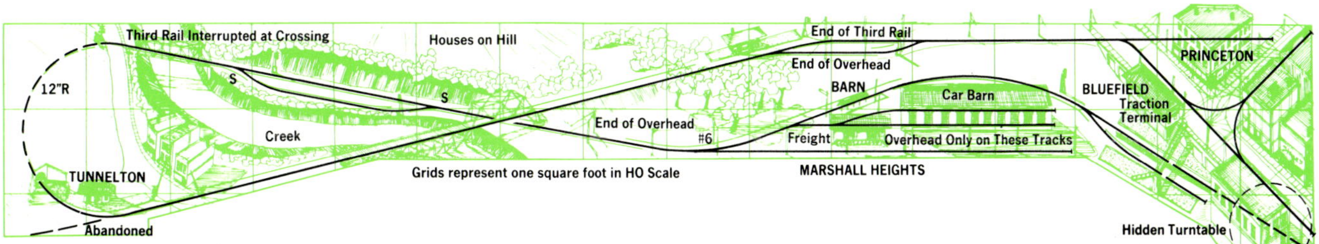

The Bluefield-Princeton Interurban mentioned on page 13 is a long, narrow interurban trolley line easily adapted to all scales. In HO it measures 16 ft. long by about 2 ft. in width. It could be built in O scale in almost the same dimensions if your trolleys can turn on 12″ radius curves. This design adapts to a variety of length and width changes in any scale. The original design called for cars to operate from both overhead trolley wire and sections equipped with outside third rail.

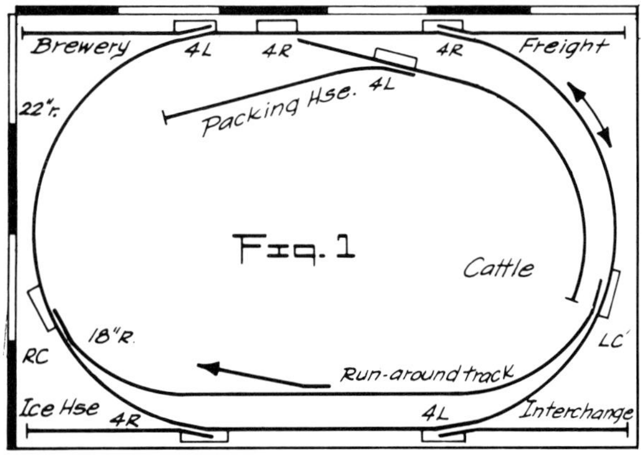

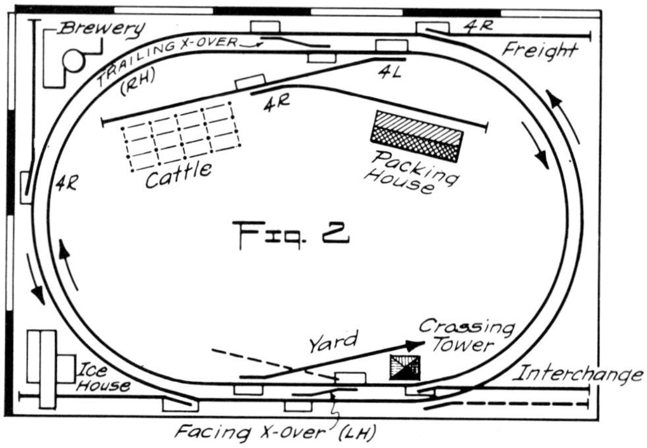

Adding Industrial Sidings

Industrial sidings are the essential ingredient for any modeler who wants to operate his railroad as the real roads do, with plenty of switching and intricate moves that make the hobby a challenge.

Industrial sidings add interest to any model railroad layout. They not only give places to locate the interesting buildings you have made—factories, mines, refineries, freight houses, etc.—but in so doing give your way-freight places to set out loaded and empty cars, and pick up empty and loaded cars a day or so later; hopper cars at mines and coal yards, tank cars at refineries and bulk plants, refrigerator cars (reefers) at packing plants and meat wholesalers, cattle cars at cattle pens and packing plants, box and flat cars almost anywhere. With the aid of remote control of direction, speed, and coupling, your "Bow-wow"—as they call the Way Freight on the Wilmington & Northern—can proceed on its route, setting out and picking up cars, giving you a lot of "operation" on the relatively small mileage of a model railroad.

Your model railroad should be designed WITH industrial sidings, even though you do not have enough switches to insert them at that time. If you do not have enough such spurs on your model pike, you can add them, preferably using one of the insulated frog unit switches so as not to upset your 2-rail wiring system. I propose now to give you some of the ideas on industrial sidings for model railroads.

Fig. 1 shows the general idea of my first 2-rail pike, the HO 2-rail Moonlight & Violins Ry. (abandoned 1940). On a 4' x 6' table, accessible from all four sides, I had an oval of maximum radius. I had a run-around track (hardly a passing track, since these must number at least two!), with home-made curved switches, somewhat as drawn. I had a yard with engine terminal next to it. And I had industrial sidings at all four corners as shown, as well as sidings inside, leading from the back. This last was feasible only because I could walk around to the back, which makes me point out this rule: Switches on model railroads should be easily accessible, within 2 to 3 ft. of the edge (preferably less), any gauge.

Since the line was single track, trains ran either way on the single iron and switches to sidings could face both ways. The counter-clockwise freight could serve the ice-house, frt. house, and packing house spurs, while the clockwise freight would serve the interchange, brewery, and cattle spurs. Actually the run-around track permitted either direction "bow-wow" to serve any siding. The extra switching moves needed to do so were actually fun, though real railroaders would doubtless consider them work.

Now had I double-tracked the same layout, as in fig. 2, I would have run into trouble because certain of the sidings would then *face* the direction of traffic. On a double-tracked line, with right-hand operation of trains assumed, no siding switches should face traffic if possible. We could call this another rule; you'll note that real roads follow it, if you'll observe. I could turn the brewery siding around, as you see. I could turn the yard around, too, as you see (although it really should be off a separate run-around or yard lead as in fig. 1). And I could and must then reverse the direction of the cattle-pen/packing-house duo of sidings. Thus the counterclockwise trains on the outer track would not hit any siding switches head-on, nor would the clockwise trains on the inner track.

But with 2 tracks, you need con-

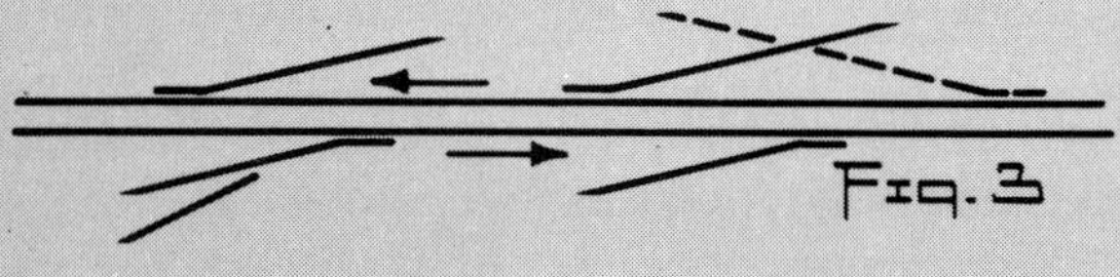

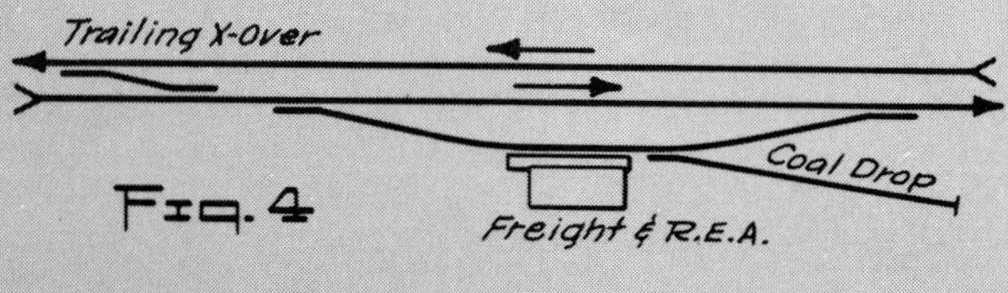

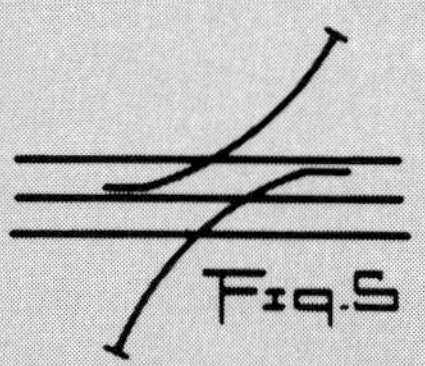

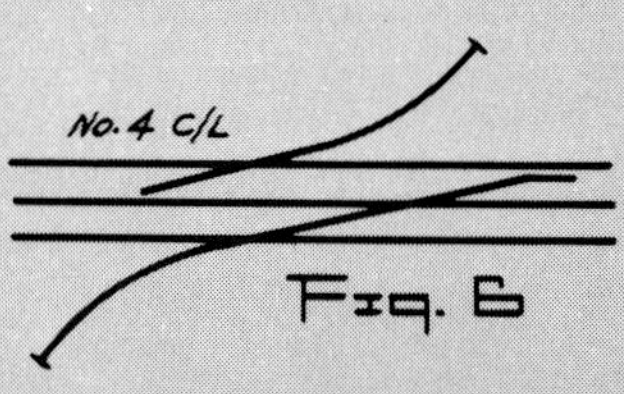

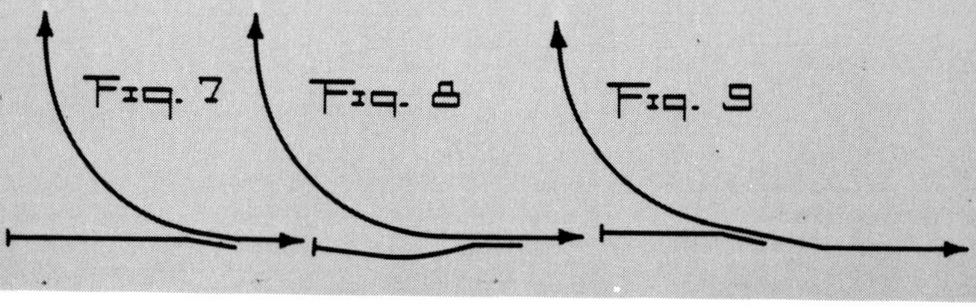

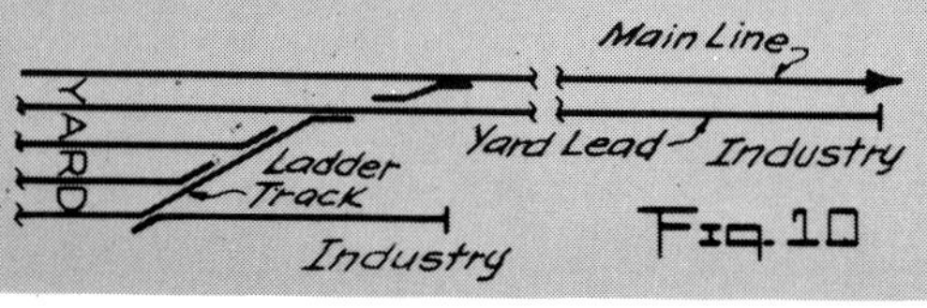

nections between them. So, I could put a trailing point (right hand) cross-over at the back, and a facing point (left-hand) crossover at the front where it could be "protected" by my hands more easily, as well as, officially, by a tower. (If I could afford only one crossover, I would make it a trailing point one—right hand—and locate it at the front).

Look, now, at the interchange track. It could be reversed, as was the brewery siding, or it could be left as a facing point siding as shown dashed, but one might be tempted to try to switch it off the inner track, as shown, with the aid of a crossing. This is NOT a very good idea since the crossing would have to be a specially made one, curved over straight. If this was to be, on your version of the road, an interchange with another railroad, you might prefer to make it a facing switch, as shown dashed: clockwise trains on the inner track could serve it by using the left-hand crossover at the front, and it would be protected also by the operator's nearness and the interlocking tower. Also, if this interchange were later extended across the chasm around the table, as I really did about 1939, a facing switch at the junction of main and branch lines would be quite in order.

Figs. 5 and 6 show why the interchange track with crossing at lower right (as in fig. 2) is not so practical, using commercial components. The arrangement in fig. 5 is fairly common where space is tight. I know I've seen in on Delaware Ave. in Philadelphia, where you had to switch from a far track, crossing the others, in order to get straight into the factories and piers close-up to the street. However, without specially made crossings, curved over straight, the modeler must use commercial crossings, #4 Custom-Line being most practical, as in fig. 6. But you see how the introduction of the straight into the sharp curve fouls up space saving.

Generally, on a double track line, all siding switches should angle off the main line track in a trailing direction. Thus, if a switch is accidentally thrown, or is not quite all the way closed, there is no danger of a train being diverted into a spur track with disastrous results. The dashed siding in fig. 3 would not be correct. Fig. 4 shows how it can be done, the old track layout at Riverton, N.J. on the Camden & Amboy (PRR), (presently reduced to single track). Since the coal yard siding, because of the shape of the building it entered, could not be made a trailing point switch, the siding was run off the freight/REA track which was made into a complete run-around track, in effect.

Thus the morning freight, running from left to right, could put the loaded hopper in front of the freight house, run around it, and push it into the coal yard. The reverse would be possible for empties being taken away. The fact that the freight crews would

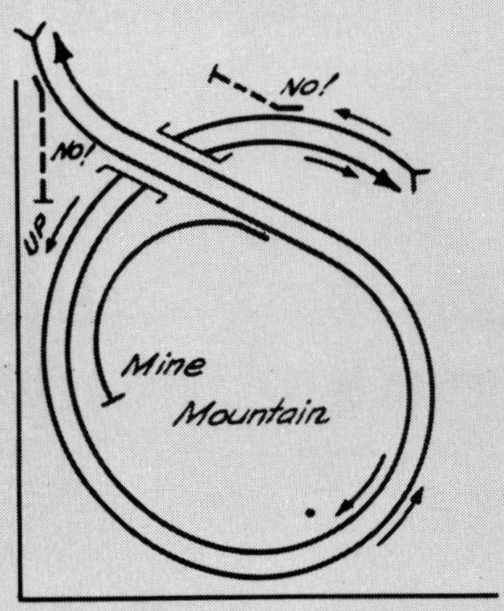

Fig. 11

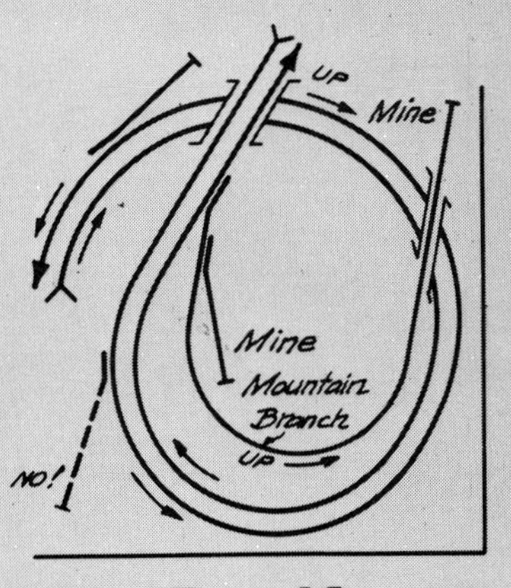

Fig. 12

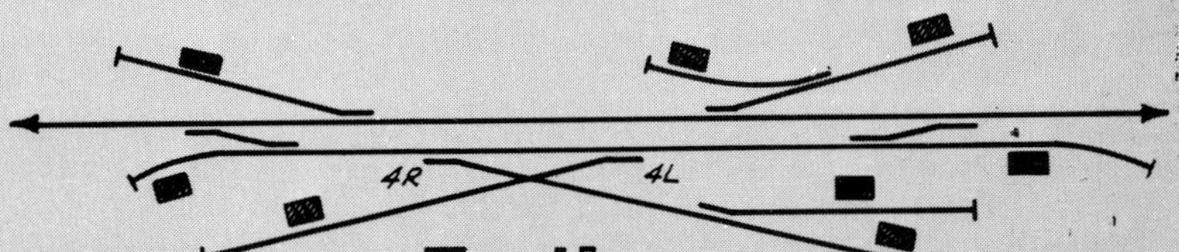

Fig. 13

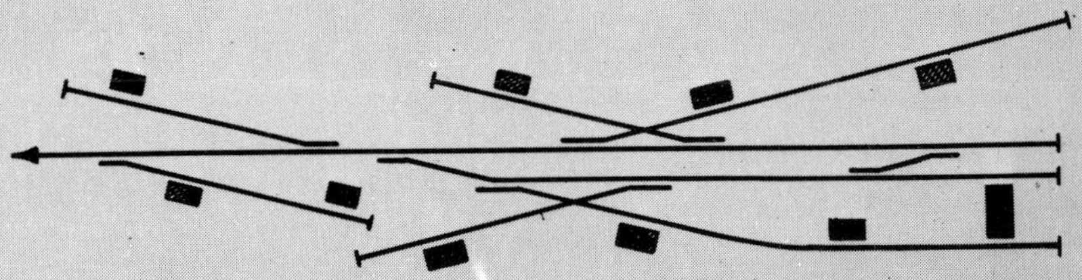

Fig. 14

actually put the car in with a flying switch, and remove it by gravity, does not lessen the importance of the layout here for a *model* railroad. We cannot make flying switches, and gravity switching is pretty iffy. So we need a run-around. The trailing crossover at left was one of many located every five or so miles along the double tracked right-of-way. But it was so located that the frt/REA siding as well as the coal drop could be served by trains proceeding from right to left! One of the afternoon trains from Trenton to Camden would stop at the crossover, then back through it to pick up a loaded REA car. I know, because I'd hop off when it stopped and save a three block walk home from the Palmyra (next) depot.

You'll note that in fig. 7 I have just drawn again the switch to a siding running straight while main line curves, utilized in the layouts fig. 1 and 2. I can't recall a single instance of this on a real railroad though doubtless some reader can. Yet it is extremely convenient on a model railroad where there are more of the right-angle curves than on real ones. For model railroads, really, either fig. 8 or 9 would be preferable IF you can spare the space. In fig. 8 the switch is ahead of the curve, while in fig. 9 the switch is on a short length of straight track inserted in the curve. Because they take more space, these better arrangements are seldom used in model railroading.

You can put industrial sidings around yards, too. Fig. 10 shows 2 very popular locations. The yard lead can be extended and the extreme end used for an industry, or, indeed, even for a branch line. (So can the other end of the lead, not detailed). You can also fill in the wasted triangle left by the usual ladder track by backing a siding off the farthest yard track, as shown.

Now the need for making switches accessible and for having trailing point switches on double track, gets you into some circumstances as shown in figs. 11 and 12. In the spiral in fig. 11, the top-of-the-grade siding must swing off in such a way that it must be very short, or sharply curved. It would be far better if the switch could face the other way, as it does when the whole spiral is located in another corner, fig. 12. However the facing switch would not be healthy. You could, however, if it were very important change the entire double track layout to left-hand running as used on the Chicago & Northwestern, and on Russ Porter's Western Valley R.R.

The Riverton layout, fig. 4, has been augmented many times in real and model practice, one such possibility being shown in fig. 13.

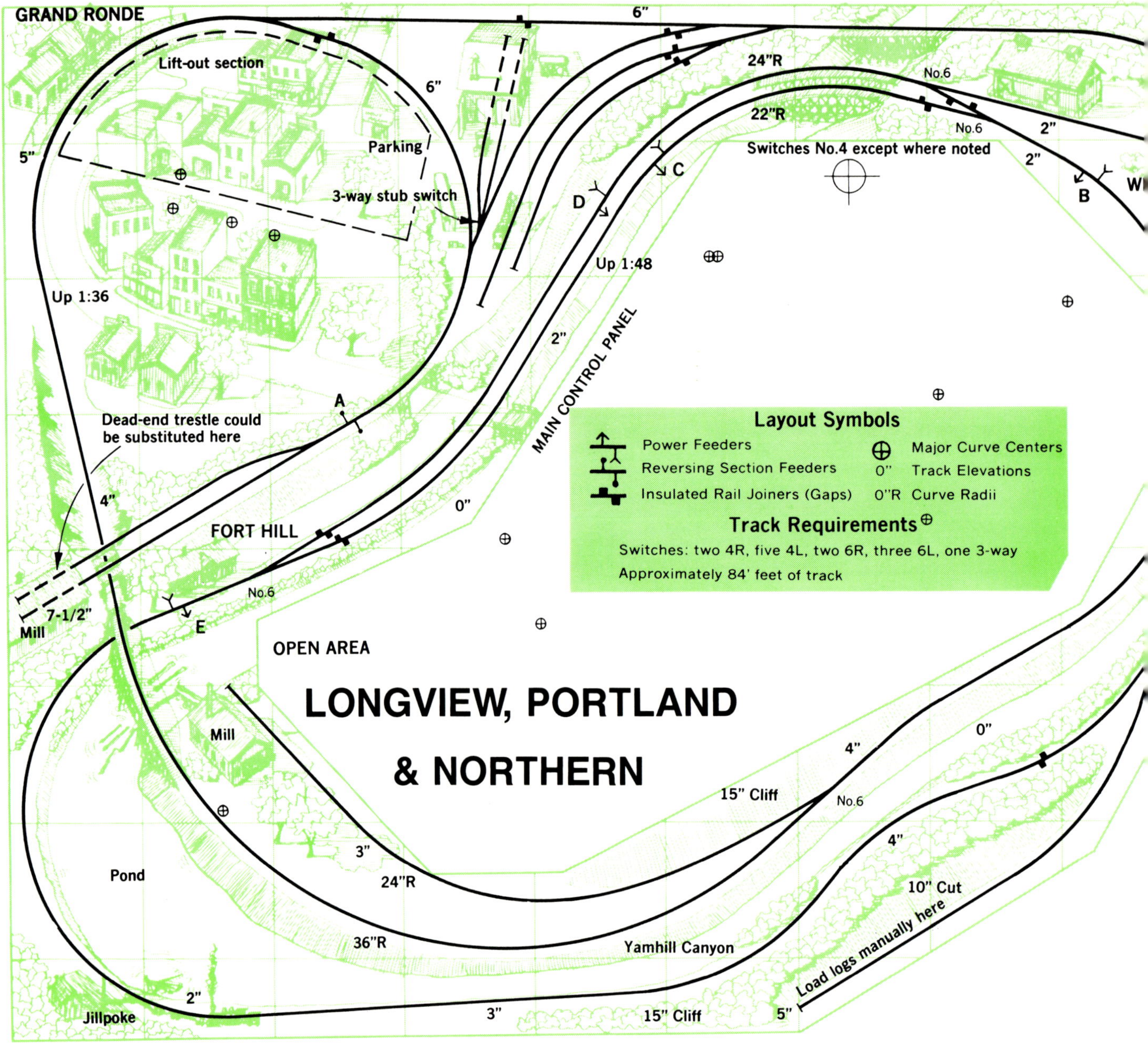

Using Wyes to Reverse Equipment

SCALING DOWN THE PROTOTYPE

Most modelers must compromise their quest for 100% scale due to space limitations. A real mile would scale from 24 ft. to 240 ft. in the popular model scales. Modelers also have to adapt larger railroad facilities to more practicable size: yards, engine and passenger terminals must be shrunk to fit the average layout. The modeler can use theatrical tricks to make a facility appear larger, using city or country background scenery to give a more spacious effect, and even using mirrors to reflect what appears to be a facility double its actual size. Paul Mallery covers such tricks in his book, (C31) Design Handbook for Model Railroads, from Carstens Publications, Inc.

Designing a logging railroad along the the lines of the Grande Ronde is not quite as easy as it sounds, even given the good ideas presented and the photos and dimensions of the buildings. It must have, of course, the qualifications of a good track plan, and some extra ones besides.

Curves should always be of reasonable radii, the minimum being the largest possible under conditions. On the GR, the range is from 18" to 36". Switches should be of the highest frog number compatible with the curves. On the GR, they are #4; there are also some #6's where they seem more appropriate. In this "backwoods" setting, I recommend using hidden switch machines. Grades should be within reason. On the GR, the maximum is 1:36 or 1" rise in 36" (the

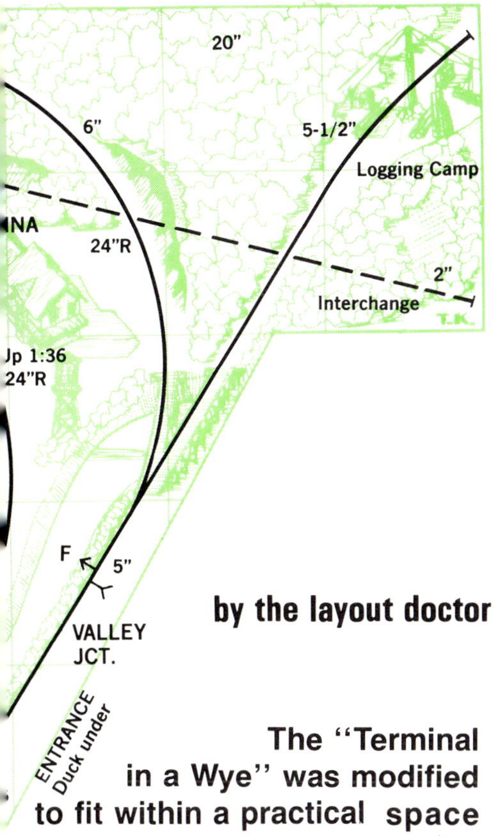

by the layout doctor

The "Terminal in a Wye" was modified to fit within a practical space when built in HO scale.

36" can be measured horizontally *or* uphill, the sine and tangent of such small angles being so close), with others being 1:48. These correspond roughly to 3% and 2% grades, reasonable for model railroads. Of course, on a logging pike, 1:24 or 4%, 1:20 or 5%, and even steeper grades can be used with geared locos or even cable, but these were not needed here.

Above all, track and especially switches should be accessible to the operator for maintenance. This means that they should not lie beyond his maximum reach, which for the sake of argument I'll define as three feet for the extreme, two feet as a general maximum, and one foot (from the edge) as best of all. These figures hold whether the layout is N or #1 gauge.

Reaching all three switches of a wye is pretty difficult to arrange except in special circumstances. On the GR layout, it has resulted in the use of the common trick of having a drop-out section in the town in the upper left corner of the plan for access to the switch there, as well as the scenery. I am not so sure that I would not, were I to build this pike, have just the frog of the switch back there, and run it around, gauntletted, with the points themselves on one or the other side of the 36" radius trestle at far left center.

A model railroad layout should have plenty of track space for storage, since cars are not constantly off-line in interchange service. Really more important, model rails usually rapidly acquire many more locomotives and cars than a railroad even twice that size could justify owning. Often trackage that is not needed in the early stages can be provided for by leaving the roadbed of "abandoned track" which is later installed, to provide more storage space. On the GR, this could be the twin tracks into the mill, at left center on the stem of the wye, letting the tracks temporarily end on one side of the trestle as on the prototype. Also, the outer or second track on the right leg of the wye is drawn with two concentric storage tracks. One or both of these could be omitted, and the weed-grown roadbed put in place for their subsequent installation. Switches could be installed for these, but spiked over.

The Grande Ronde is prototypically a straight line railroad, running from the above-mentioned town to Willamina. This could be duplicated with a point-to-point trackplan in a small space. Many modelers have found the slow pace of such a pike to be ideal, and the lack of a continuous-running provision is offset by the fact that it takes a long time to run between the two terminals due to speed restrictions (poor track) and the need for the train to stop at each and every industry and station. A hot shot through train has little place on this railroad, but it does provide a nice setting for small Moguls, Prairies and the like. Too many modelers are too ambitious in planning their first layout, and basing the pike on a smaller prototype might provide a partial solution.

However, for test running, break-in and a chance to simply watch the trains roll on occasion, I have provided a design offering continuous running for the Grande Ronde.

In spite of the praises of Dave Lewis in the prototype GR article, a terminal for engines inside a wye doesn't always result in the advantages for *model* railroading he suggests. Real railroads usually have a much greater space in which to spread their tracks than does a model railroad; hence, a wye or any other special trackage on a real line can s-p-r-a-w-l, whereas on a model line there are always good solid walls or precipitous table edges to restrict curves, etc..

A study of a diagram of the prototype indicates that the minimum radius of the wye tracks would be (for HO) 22", with the optimum somewhat larger, 30" or 36". If ready-to-run trackage is to be used (which as noted above I don't recommend in this case), 22"R it must be, since the only HO curved switches available have 18"R springing from the inside of 22"R, needed for the "old engine house."

Many preliminary sketches were made before drawing the "final" wye arrangement shown here, utilizing a three-way switch (which could be a stub rather than point design on a pike of this nature). I have used fairly standard curve radii and switch frog numbers for those who insist on using sectional track components. However, hand laid track would be ideal on the GR, and the Kadee Rail Spiker is a very handy tool to speed up this work with no loss of realism. Don't overlook switches offered by firms like Simpson Products in various rail sizes; these can save much time when many turnouts must be fabricated and are of excellent quality. Longer pieces of flexible track and turnouts with code 70 rail are almost impossible to distinguish from hand laid track after ballast, weathering and weeds have been applied. In short, super-detailed track is not hard to install on any layout with available commercial components, especially in HO, S and O scales.

The design of a logging pike has one additional qualification that influenced this trackplan: tunnels are "out", though I'm not prepared to make a flat statement that tunnels could not be found on logging railroads. However, imagine what would happen to a log train if a log or two came loose or even shifted a little while inside a tunnel. There have been tunnels on logging roads,

Yard and terminal areas are easily cut from 4x8 sheets of 5/8" plywood topped with Homosote to deaden the sound. A sabre saw is used to accurately cut the turntable pit. A portable circular saw is best for cutting the 4x8 sheets to the required size. Many small layouts using sectional track are build on such a base in their entirety.

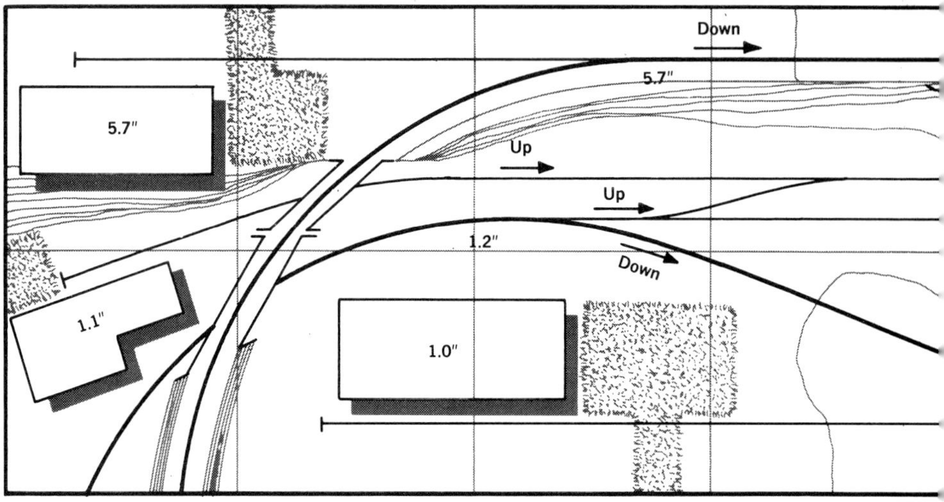

by Al Westerfield

The elegant back door: a "paired industries" trackplan

Back-to-back modules form the basis of a logical and compact track arrangement for industries which share loads and empties.

A "back door" layout is one featuring hidden connections between tracks that appear to be dead ends. The most useful application of this type of plan allows consists to always travel to their proper destinations. An example of this would be an ore pit to smelter run. Empty ore cars are delivered to the diggings; loaded ones are hauled to the smelter where the process starts over again. Any load that requires an open car—coal, gondola or flat car—looks best when following this prototype cycle.

The only requirements for building a back door are enough siding space to drop two strings of cars (any length) and the cars themselves. There is one problem, though: it always seems that a back door takes a lot of space if there isn't much to spare, but relatively little on a large layout. Fig. 1 can help explain this paradox. Here we have the simplest form of the plan, a main line and passing sidings. The sidings are visually bisected in some manner, giving the impression that they stop at the industries. The tracks could run through a mountain or simply through a wall or dividing board, *but this fact is camouflaged*: the tracks run into a building, behind a rock outcropping, etc. The greater the visual distinction between the two scenes, the more realistic the effect. This is easy to achieve on a large multi-level layout where the connection can dip down under other railroad scenes and emerge 15-20 feet away, much like John Armstrong outlined in *Model Railroader*; I'm building one like this on my John Galt Line. But the smaller layout will require a much shorter and direct connection, yet still be long enough to take a string of cars.

The operating mechanics are simple. Referring again to Fig. 1, cars are backed into one industrial lead and emerge for pick-up at the other industry. The loco drops these and picks up the cars on the other track, pulls out onto the main, and delivers them via the visual route. The same procedure takes place upon arrival at the second industry. On a large layout, the cars might be interchanged at a yard during delivery. This is impractical where space is at a premium, so our plan must be refined somewhat.

Fig. 2 adds diagonal track to the circle, forming a double loop, familiar to anyone who started with a train set. This design permits a loco to back cars into either industry without run-arounds. But unless the back door is to be the main feature on the layout, this plan obviously takes up too much space. Compression and elongation give a dramatic new look, Fig. 3. Here we see exactly what our back door really is: an elongated crossover. Assuming right-hand running (deliveries as shown in Fig. 2), if both rails of both tracks are gapped as shown, polarity problems commonly associated with loops need not concern us; the loco will never run over the gap in normal operation.

The dog-bone with crossover is the basic form for any small back-door layout. A lot of variations are possible, of course. Fig. 4 is one example, a small around-the-room design. By putting both loops of the dog bone in one corner, only one area need be as wide as the minimum radius. The remainder can be a shelf or other space-saving design.

The back door trackplan shown here is an amplification of Fig. 4. It is designed primarily for the length of a small room and HO scale. All grades are 2.5%; change of grade is indicated by an arrow next to the tracks; minimum radius is 24". Obviously, the plan is simple enough to be used in larger or somewhat smaller spaces by either straightening the track at each end or shortening the back door tracks. The back door as drawn provides room for eight or nine twin hopper cars, a good train length for the size layout envisioned. There is no law that says the train has to be unitized, however. Small facilities might only require two or three cars. What I'm saying is that this plan should act as a catalyst for further design to suit your own space requirements.

The plan is divided into two modules which I'll call City and Mine. The meeting point is quite naturally at the scenic divider. I suggest that this divider be a solid sheet at least as high as eye level and extend from the backdrop to the table front. This makes it impossible for someone to see both modules simultaneously at close quarters. The result is a visual isolation which fools the viewer into thinking that more than a few inches exist between modules.

The City module is designed for as much operation as possible. The central feature is, of course, the power plant. The double-tracked spur comes off the main on level ground and immediately begins a 2.5% grade. The tracks level off just outside the power plant, giving the impression that they end in the plant. Actually, the grade begins again once inside. The plant itself should be massive in appearance. The structure could probably be kitbashed from Vollmer, AHM, Con-Cor, MDC, etc.,

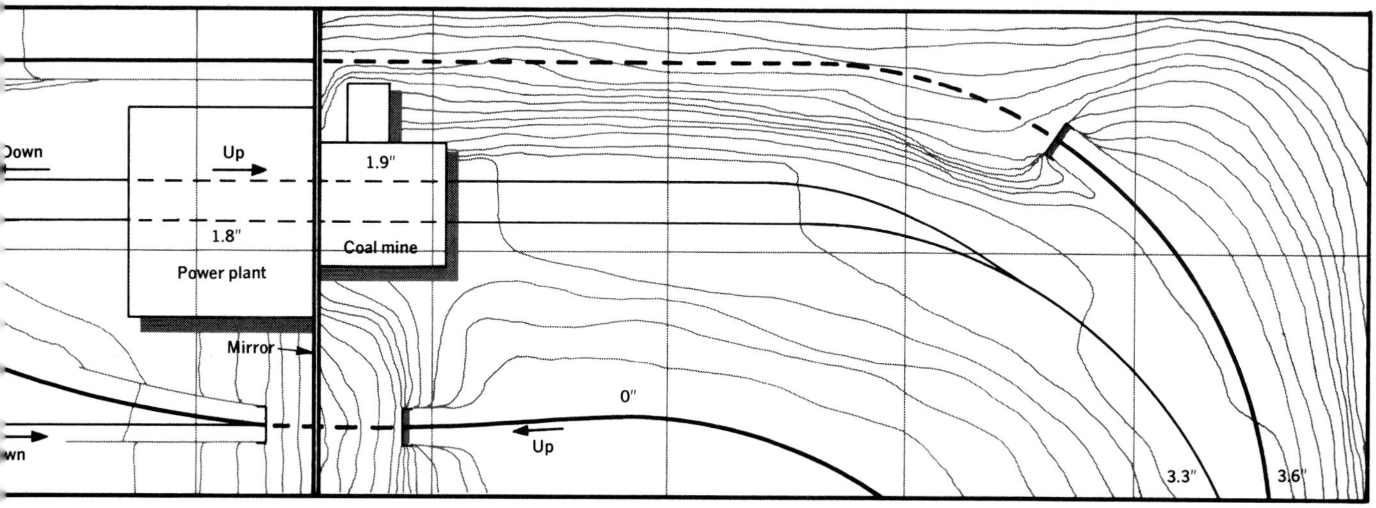

Artwork: the author

plastic kits. For best effect, the building should be an extended facade mounted flush with the divider, and the divider should be a mirror. Not only will the building seem twice as long, but the entire city will then extend in both directions. Notice that in no case does a track run directly into the mirror; "visual cornfields" are scrupulously avoided. The rear track is obscured by the power plant, while the front track disappears into a subway-type tunnel.

The most important effect to duplicate in the City module is the feeling of claustrophobia. Land is always at a premium, so crowding is the rule. A facade of tall buildings should dominate the backdrop and overwhelm the trackage. Trackside areas should be dirty and neglected. This is best simulated by avoiding strong colors, especially the typical "shocking green" lichen. Litter the landscape up a bit. Give a coating of grime to all structures. More instructively, study all the city photos you can find of John Allen's layout ; in my opinion, a realistic city is the most difficult scene to model, and John's was the best.

The Mine module portrays almost the exact opposite effect. Here nature reigns almost supreme, marred only be the coal mine. The rear track is hidden for most of its journey to better give an uncluttered feeling to the mine area. With the tracks at three different levels, a further spacial distinction is made by the viewer. If possible, scenery should be extended both above and below track level to enhance the effect of wide open spaces. Strong colors here will look healthy, vital.

The mine should be built onto the side of the mountain with the loading tracks running underneath. Strategic placement will hide the backdoor from all but the snoopiest visitors. Movable steel wires which can be projected up between the rails catch the axles of the hopper cars and prevent them from rolling downgrade. The front and back tracks run into the mountain through the more conventional tunnel portals.

These two modules illustrate the basic back door. Using the basic rules and a little doodling, I'm sure you can design one for your space. I think it's worth the effort.

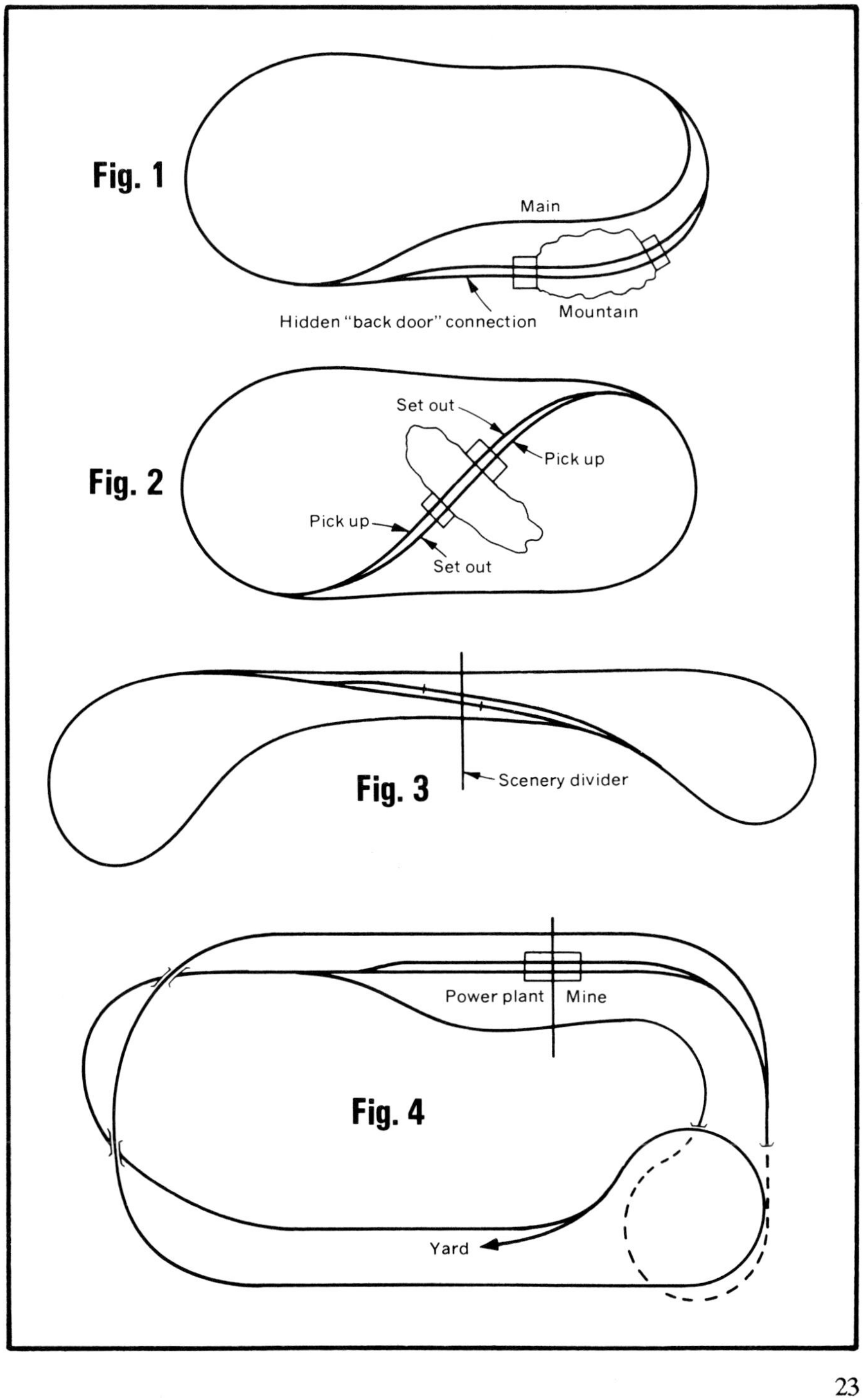

One-horse Short Line

Sharp curves, small yards, one loco — a formula for modeling fun.

by the layout doctor

Above photo by E.H. Blabey; below Richard Allen

Prior to a keen interest in heavy main-line railroading, even prior to a fondness toward interurban lines, the type of railroading that held top spot in my affections was backwoods, one (iron) horse lines. Such lines once abounded, feeding a little traffic to the big trunk lines, sometimes being former branch lines thereof (sometimes even future ones). If you can locate a railroad map of any state, made in the 'twenties or before, you'll find them. If you seek to trace their routes out in actuality, you'll often find only parts of them near the main lines still in use for industrial purposes, but otherwise just ill-defined rights-of-way, abandoned abutments, or maybe a super-highway route. Some are still very much alive, of course.

Motive power on the one-horse short line would, likely as not, be a small cast-off from a nearby steam road (often later recaptured and preserved by the big road!). Baldwin 4-4-0's were common, as were straight stacked 2-6-2's, 2-6-0's, 4-6-0's and an occasional 2-8-0, most with small drivers and small boilers — all lightweights because of the puny track, poorly kept roadbed and shaky bridges, and held down well below top speed. You might even find a geared engine in some areas, or perhaps a gasoline powered rail-bus or automobile on flanged wheels. Company-owned cars would be limited to a few flats for a work-train, and a combination car (with open platforms!) used both as caboose and passenger car on the daily mixed train — if, indeed, service was *that* frequent.

The Livonia, Avon & Lakeville in New York is typical of scores of short line railroads that are fun to model. Their small steam locomotives were often bought second hand from larger railroads and they sometimes offered mixed train service. Scheduled passenger service might have been a gas-electric or a small locomotive with one or two passenger cars. Short lines are great fun for modelers who don't want lots of equipment.

The right-of-way might be plain dirt ballast, but more likely it would once have been built to high standards and given no more than a lick and a promise since. Weeds and shrubbery would abound except where burnt off by a dropped hot clinker or spark from a defective stack.

I've drawn up my idea of such a branch — really an independent short line railroad — herewith, admittedly based on the idea that you will not be satisfied with just a daily mixed but want a few more action possibilities.

The Wagner County Railroad is in reality a continuous-run oval filling both 4'x8' lobes appended to a 3'x12' panel. Dimensions are for HO, as is most of this commentary. As usual, if you want to build it in N scale, halve all lineal dimensions; if in O scale, double the same dimensions.

The junction with the main line is at Wagner, where the Wagner County crosses the main. The WCR is really two branches, one on each side of the main. The main

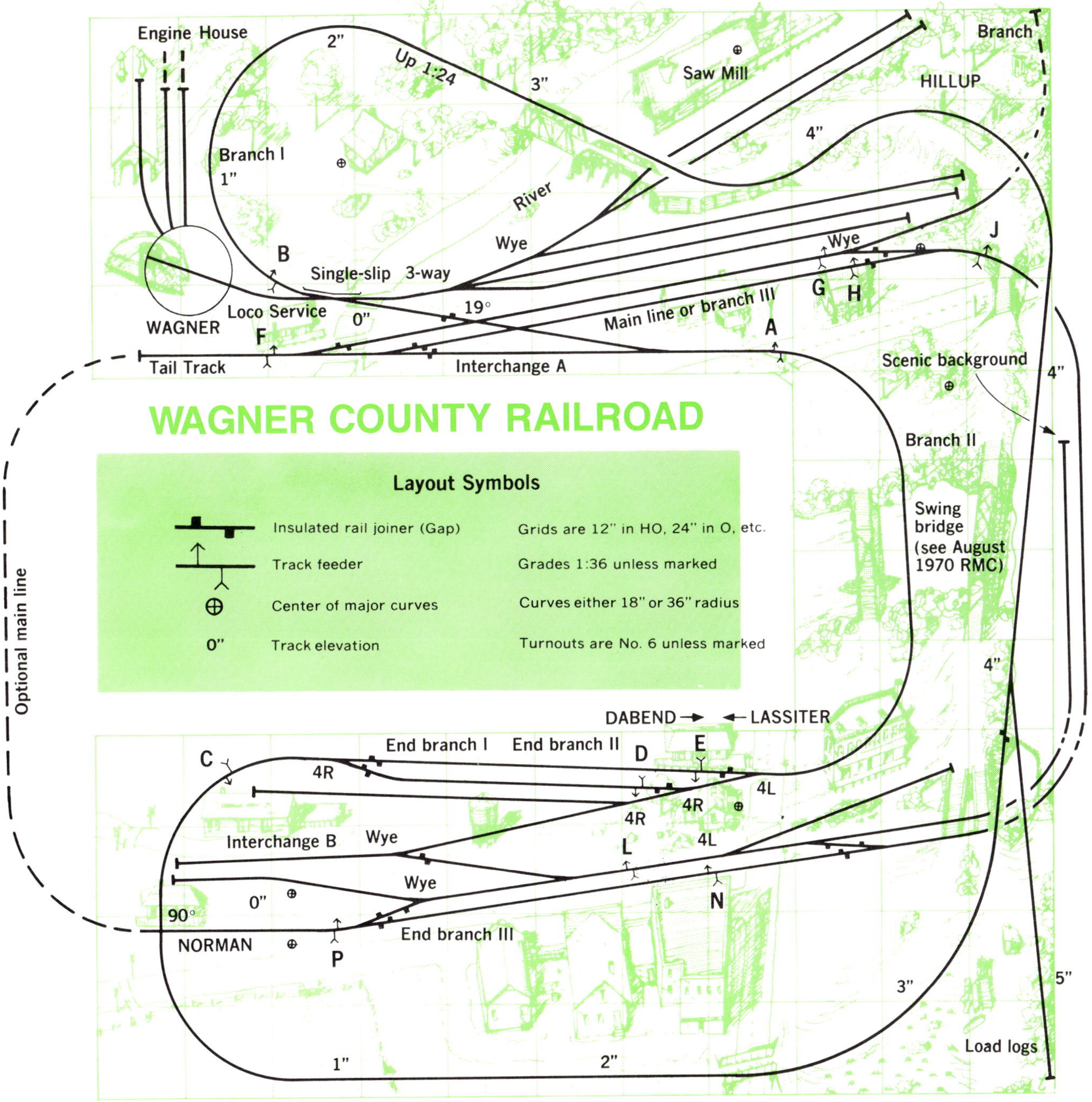

line could be continued into new territory, as hinted by the broken lines, for continuous run. Alternately, it could be treated at times as a third branch of the WCR.

Just picture this: three trains lined up at Wagner, all scheduled to leave at the same time! Train A would be behind the station, ready to head up the grade, ending its run at Dabend on the other panel. Train B would extend from the interchange track toward the station, and would leave via the drawbridge for Lassiter. Train C would be to the right of the station, blocking the crossings, and would leave via the main line for Norman, which really isn't very far from Dabend/Lassiter.

Notice that the grade up out of the yard is pretty steep, even though it could be gentler for there is enough space. This is so that switching in the three-track yard can be done by gravity, down the 1:24 (4%) grade. The approaching line has been ducked toward the front for accessibility so that you can boss the uncoupling of cars to be switched in case the automatic uncouplers do not work perfectly. From here the cars will coast down the 1:24 grade (with a guard rail against inner rail) and into the yard, something like having your own little hump yard.

At Wagner there is a turntable and a two stall engine-house with tracks run on into the hill so that each should hold two or three locomotives. There is also one outside track. While some purists would maintain that a one-horse line should have but one locomotive, most of them will allow *at least* one spare.

Another operational idea is to have the swing bridge at right center out of service, either not built yet or just plumb worn out. This would let you run trains from one shore, all the way around across the high bridge to the other shore. These stub tracks to the bridge could then serve industries.

For sake of appearance, track should probably be made using individual ties and either code 70 or code 40 rail instead of the usual code 100 (excellent code 70 track and switches are commercially available). Rails should be painted rust color, with head polished bright for contact. And there should be plenty of grass and foliage in and around the track. You really should try to keep the track relatively even and smooth, for model cars will not take the rough track that real ones will: they have deeper flanges, but not the great weight.

Track parts required for layout built as drawn (without optional track at left) work out as follows: one #6R switch; four #6L; three Y; one single (or double) slipswitch; six #4R; three #4L; one 3-way; one 19° crossing; one 90° crossing; 150' of flexible track. These requirements would be the same for N or O gauge except that the feet of track would be halved or doubled.

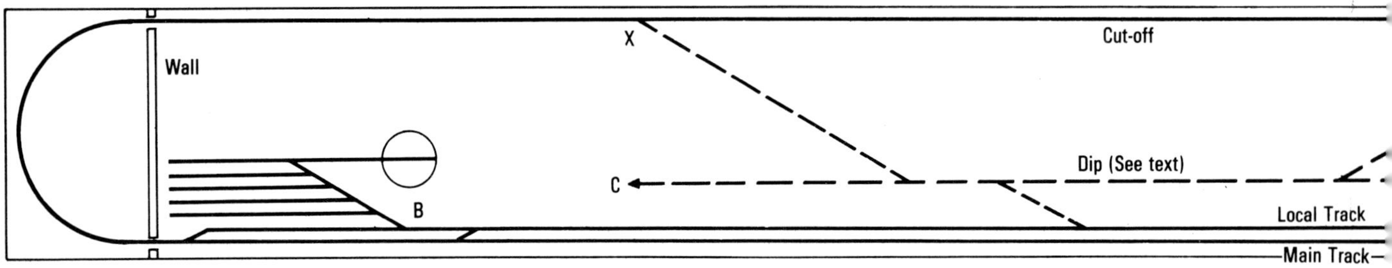

LONG GUYLAND RAIL ROAD

Long, Lean and Lanky Layouts

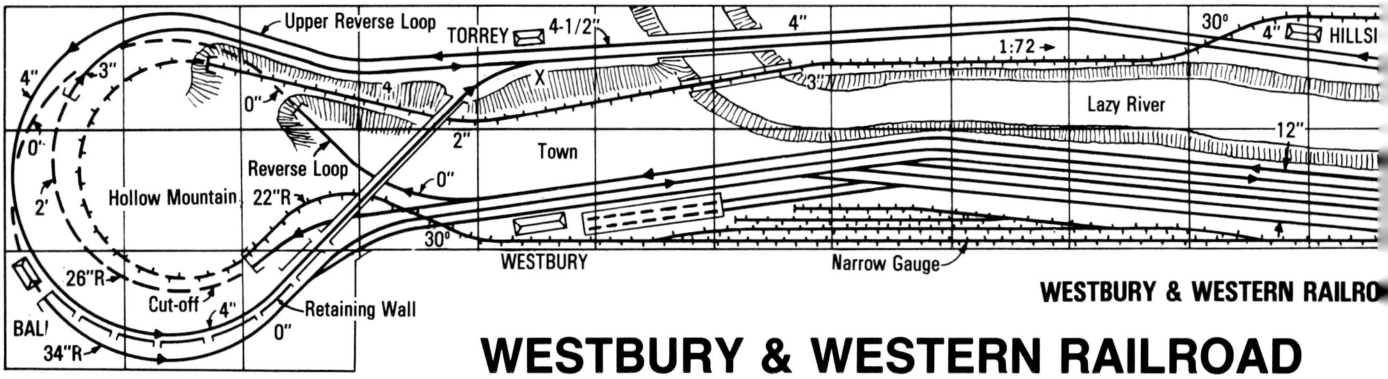

WESTBURY & WESTERN RAILROAD

These two long, narrow layout designs can be adapted to any length, and provide some interesting possibilities in any scale along with a few minor problems

If basements are a rarity in many western areas, they are common enough in the east and midwest. With the advent of ranch houses more and more basements are turning up that measure 50 ft. and more in length. Not surprisingly, some modelers are pondering new layouts that are lean, long, and lanky; meaning that they are getting the bug for modeling a stretch of railroad and is prototypically straight with only a few minor sweeps back and forth to make the scenery interesting. At each end, of course, is the same kind of loop you'd find on almost any circular pike—a full 180°.

A tinplater, and this might be doing an injustice, might be tempted to race his trains up and down the straightaways. Hopefully, our readers have matured enough to know that trains start slowly and proceed at prototypical speeds. A long freight of 1920 vintage might be lucky to creep along at 20mph. A modern day manifest freight might do 50, 60, or more. Even passenger trains don't normally move at 100 mph. Passenger trains might occasionally hit 80, 90, or 100; but chances are their average speed is more like 50 or even 40mph.

What is remarkable in a layout such as we show here, drawn 40 ft. long and only 4 ft. wide excepting for the return loop areas at each end, is the total length of the run: 1-1/3 miles in HO and a staggering 2-1/2 scale miles in N.

The layouts are also highly versatile. The modeler can shorten them to any needed length, perhaps 30 ft. long and only 3 ft. wide. They can also be expanded to 50 or 60 ft. in length. It wouldn't be any great problem to bend the layout into an el shape almost anywhere in the middle.

The *Long Guyland Rail Road* is presented as a schamtic diagram, more than as a finished layout. It points out one question, important on any pike, but critical on a long, lean, narrow layout: where will be the principle control location? If you want it at Terminal A, fine, but it is a long walk to the other obvious control location at Terminal B. Duplicate controls are one possibility, having a Terminal B control back at Terminal A, *if* trains cooperated by always staying on the track, never derailing or stalling, and always coupling and uncoupling 100% of the time. You could plan for two operators, but rarely can the home operator be assured that the second man will be available on operating night.

You can of course build the Long Guyland with just one terminal, A. But if you do want two terminals, the idea shown is the best one. You could "dip" the back line forward as shown dashed and move the second terminal to a site such as D, next to A, where its switches would be reachable from the same site. A terminal such as C would have to be backed into by a train coming out of A, and so would not be as useful.

I suggest that the back track be "dipped" forward to make the switches more reachable. The two switches marked X, at each end of the "cut-off" at rear, would be four feet back from the edge, barely possible to reach. Hence the cut-off would have to go and the track would have to be dipped forward as sketched.

Another idea is the wall across one end of the pike. This could be a "natural" wall, tunneled through, but it could also be built up and covered with suitable scenery. It would get the track out of sight for at least one portion of a train's trip. The hidden part at left could even be in a tunnel under a mountain instead of just behind a wall.

Yet another idea, usable on other layouts, came to me as I sketched this out: that is to have double track along the front, not just as keep-to-the-right double iron, but as two separate lines. The outer one would be the main line; the inner one would carry rapid transit electric (or diesel push-pull) trains from A to B and return.

Notice that the right end, near terminal A, has been widened out from four to six feet, always a good idea if you can manage to get the space. You could even make

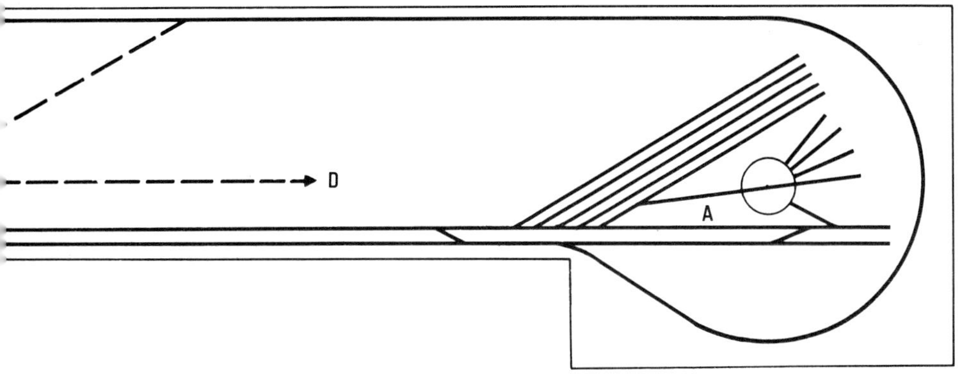

Size of Squares: 24" for HO or N

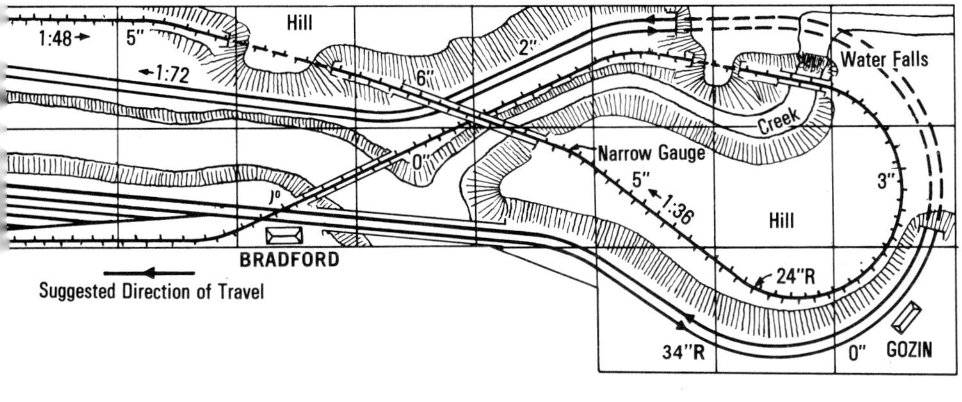

by the layout doctor

the "neck" of the layout narrower than four feet if you could swing the curves at each end to a wider, easier radius.

So far (and for the rest of the article, too) I've been thinking and talking about either HO or N gauge. While any HO layout can be built *the same size,* instead of half the size, in N gauge with suitable minor redrawing, a long, lean, and lanky pike makes such loose talk much easier because the layouts can be built almost any size.

Such layouts are hard to draw because, unless drawn to a small scale, they are almost impossible to take in at one or two glances. With a pike 4x6 or 8x10 ft. in size, you can take in the general idea with one good look, and then search out the details. However, if that pike has one dimension of, say, thirty feet, you must move the plan around, or roll and unroll it like a scroll, making it difficult to comprehend without checking back and forth.

To make it able to be taken in easily, a long pike should be drawn to a small scale. When I or anyone draws a layout plan for RMC, we do it at least *double* the size it will be finally reproduced to get better detail and iron out little irregularities. Actually I draw them to a size suitable for use as a working drawing when building a pike—for most layouts to 1" scale, or if not too large to 1-1/2" scale. The drawings, if reduced by one half for reproduction, come out to 1/2" or 3/4" scale.

Even if the problems of reproducing such a plan in RMC were not considered, there would still be your problem of drawing the plan to a suitable scale for a working drawing. A forty foot layout to 1" scale would be over 40" long, to 1-1/2" scale, over 60" long. Is your drawing board that long? Or would you have to let the paper curl off both sides of your drawing board? Or later fold it into a "book" like a steam loco erection drawing?

My solution is to draw the long, lean, and lanky layout plan to a smaller than customary scale to be determined by the drawing board at hand and, to some extent, the paper. My drawing board, the most used one anyhow, is 24" long, so that if a forty foot long layout were drawn to 1/2" scale, it would be 20" long and fit the board nicely. Within the half inch which represents one real foot, you can easily fit six parallel HO tracks if you measure from center lines of the outer tracks, five if you measure over the roadbeds. You can get more N gauge tracks than that—ten on 1-1/4" centers with a little left over.

On the *Westbury & Western R.R.*, 40'x6', with a 4' neck, I've solved the problem of where to put the main terminal by stretching it out along the center of the table. The main controls would be by Westbury station, since the right end of the yard is principally for exit to the main line. The double track main line—on 2" centers if HO, on 1-1/4" centers if N—hairpins along the front and back of the table, connecting the lower reverse loop with the upper reverse loop. There is a cut-off track connecting the inner track at high and low levels for continuous run without looping. Really, they are not electrical loops at all and no special wiring provision needs to be made (note there are no crossovers between the double tracks). All trains enter the yard from the left or Westbury end, and leave via the right or Bradford end.

The long straightaways shown in the Long Guyland are broken up somewhat on the Westbury & Western by the way the tracks dip away from the edges—at front to get around the yard, at rear to get around a hill. Note they are scenically separated by a big river approximately one actual foot wide (when not in flood—now there's an idea—a railroad running through a flooded region!).

One of the little ideas I threw in on the W&W is a narrow gauge road circling the area. This runs along the very front of the table, crosses the main at right, climbs to a higher elevation, crosses the mains again at center rear, and, after coming down to ground level, crosses the three main tracks at far left to get back to its starting point. Not too much yard trackage and sidings on this line, but it could be given more by adding small auxiliary tables at the front.

My original thought was that this would be an HOn2-1/2 line, but I soon got into a cul-de-sac when I discovered that the only HO/HOn2-1/2 crossing offered was 30°, left-hand at that. This would work at Bradford and Hillside, but at Westbury there are three similar crossings where right-hand would be required. You or Sattler of Cincinnati could make 'em, but if you are having crossings made, why not HOn3 for all and be able to operate some of your HOn3 locos on a long pike? A third possibility would be to make this an HO trolley line and include one or two interchange tracks at the front. Just picture three steeple cabs and a flat motor hauling a thirty car railroad freight train over this steep trackage when the railroad mainline is blocked! Because tracks have been drawn farther apart than is customary, you'd find in building this or a similar pike, that you'll have more space than the drawing indicates. I've already explained how you could fit six HO or ten N tracks in a 12" wide (CL to CL) right of way, such as the width of the yard marked at right center. Instead of the two main and four yard tracks drawn, you could fit in ten N gauge tracks! No. 6 switches are assumed all over, and grades as well as principle curves are marked.

I have not marked insulated rail joiners and track feeders on either of these long plans. They would be reproduced too small for the insulated joiners to show up clearly; in fact, even the original drawing would be somewhat crowded with them. If you want this information, have a photostat made of your choice of plan much larger than reproduced (four times as long, preferably) or redraw it with whatever changes you want, send it to me c/o RMC with an SAE (no mailing tubes, please!) and I'll mark the insulated joints, etc., and send it back to you.

Nor have I given specifications as to how many feet of track, how many switches, etc., are needed. The Long Guyland is only a sketch or diagram, not a finished layout plan. The Westbury & Western is subject to so much immediate revision as to gauge, size, etc., that any tabulation at this point doesn't seem worthwhile.

The Big 8 Railroad

by the layout doctor

A moderately sized pike originally designed for a wheelchair patient is high on operating enjoyment and low on maintenance problems.

This pike was designed for a specific individual. While in itself it may not be of great interest as a track plan, it should serve to point out various lessons of layout planning as well as how, at times, one has to make the best of bad circumstances. It appears that a space of over 6'-0" by 10'-0", accessible from all sides, would be easy to plan an HO pike to fit in. The hitch was that the pike was to be built for a lad in a wheelchair so that he could operate it by himself. Thus, the control "pit" was put in the front, which of course could not have a bridge across the entrance. The two 4 foot wide front lobes of the pike, then, limited maximum radius to 22" and, if double tracked, to 20"R or less for the inner track. To minimize running problems #6 or larger switchers were used.

To begin with, I suggested that the top of the main table be made at the navel height of whomever was to be seated in the wheelchair. If an adult were in the chair, it would be a little higher than if he were seated in an ordinary straight chair. This height would enable the operator to get into the control pit, chaired, and slip his knees under the railroad table framework until the arms of the chair fouled the table.

Notice how the control pit was not made rectangular, which would have inhibited the layout design even more. It is usually quite in order to slice diagonally across one, or even both concave corners of such a pit. Thus, I was able to move the double tracks at left upward so that a concentric switching lead of 24" radius could be located outside the main line.

While the main line featured 22" and 20" radii, its rear left corner was drawn with 36" and 34" curves for variety and appearance. Notice that all switches but one were kept within reach of whoever was in the control pit. The exception was the spur at extreme left, which could have been omitted, spiked straight, or just kept straight while the single operator was running trains alone. After all, since double tracks in the U.S. run "keep to the right" (with mighty few exceptions), a train approaching this switch, coming down the hill, would push right through it even if it were set for the spur. A few trucks *might* get derailed on light cars, but no grand disaster.

Switches were drawn with top-mounted switch machines in mind, which meant either Custom-Line or Casadio in HO (so far). The double slipswitch (these *do* work, and nicely!), the 3-way switch, and the two curved switches had to be Casadio (unless they were custom made to exact specifications). Casadio switches have rather deep flangeways to suit European equipment, chiefly Rivarossi-made. Thus, this imported equipment will run best through them. However, even engines and cars with NMRA flanges, which means domestic and Japanese production, will run through without real trouble—the worst being a jounce that real people would find frightening, but no complaints have so far been heard from the HO Little People. Watch out for any locos with only two wheels picking up power (0-4-0's, some B-B diesels, 4-4-0's, 4-4-2's) because they are likely to stall when running slowly. Even some larger locos may stall at slow speed due to the isolated or dead frogs, dirt, or poor wheel/rail contact. I also suggested that the left-hand switch close to feeders C on the plan be Casadio because its smaller motor housing lets the tracks nestle closer together. Other switches could have been Custom-Line or whatever: there is no harm in mixing the breeds. However, if all-rail frog switches, not meant for attached switch motors, were used, some relocation of the lettered track feeders would have been needed.

Notice that there were only two crossovers between the main line tracks, the facing or LH crossover being to the left of the main station, the trailing or RH crossover being to the right of the pit (up on the hill). The latter was to be made of two RH curved switches, 18"R springing from the inside of 22"R. I didn't like to introduce 18"R on the main line, but what else would have fitted here under the circumstances? I'd have preferred to put both main line crossovers in front of the terminal, but one would have had to have been too far back.

There were seemingly three crossovers between the "outer" main line and the switching lead. One involved the double slip and could be called a RH crossover. The exit crossover to the same main line track was too close to it, across from the station, where it could be easily reached. I'd much rather have put it back farther, under the bridge, where a crossover made of one LH, one RH switch was shown in dashed lines.

Industrial sidings were not too plentiful on the pike as drawn. The tip of the yard lead, lower left, might have had a small industry with the penalty of shorting the effective length of lead. There was a spur at far left; another, not even indicated, could swing off another RH switch opposite this, in front of the small station. I did show a LH switch, with branch shown as a broken line near the control pit and the main station. This could be developed into a tree of industrial sidings in the town area. A LH switch at center right on the outer main line could run as a straight spur, parallel with the edge, into the open area left by the inward dimpling of the mains.

The branch line (section D) was kept simple, just one siding, within reach of the control pit (branch loco waits while main line train sets off cars on the branch, where the waiting loco pushes them up the grade). With more mobility on the part of the operator, additional spurs could have been put in at far right and center top, pointing downhill. The upper end of the branch could have been divided into two tracks and could even have been elaborated into a small terminal with even more tracks. Branch curves were drawn 16"R, making them suitable mainly for small road locos or double-truck geared locos—Shay, Heisler, Climax, Vulcan Duplex. If the main line railroad were diesel, as I recommended in the particular situation this pike was designed for (since they are often less troublesome in regards running gear and keeping them on the track), your "only" steam loco could be operated up the branch, with suitable water tanks, coaling facilities, ash pit, sand house, and even some open fan-trip excursion cars.

It should be noted that the 3-way switch to the enginehouse was located under the long bridge (and so was the optional crossover from yard lead to main line); hence, these switches must be installed and perfected as to operation before the bridge is put in place. If you don't want or can't get a 3-way switch, the outside track by the enginehouse could have a RH switch swinging off the yard body track nearest it. It would be a good idea to have the bridge and its tracks removable in case much maintenance is required on the switches under it, which is why I had the figure-8 line rise to cross over itself with 4" clearance (really rail-head to rail-head measurement; clearance would be somewhat less), making it easier to work under the bridge if necessary. To make the bridge removable, have the rail joints somewhat back from the abutment at each end with rail joiners installed so they can slide back, one way or the other, and the bridge lifted out. I'd suggest making this bridge a spectacular through-truss design.

Grades on the line, which I called the Big Eight R.R. (the herald was suggested as a giant 8 with an R in each of the loops!), were drawn with 1" rise in 36" (1/4" in 9") which works out to 3 in 108, approximately 3%. This could be reduced to 1:48 (1/4" in 12") or 3 in 96, approximately 2%, if the bridge "clearance" was reduced to 3" or a fraction more. Any grade will cut down on the number of cars an engine will pull up it, 3% more than 2% to be sure.

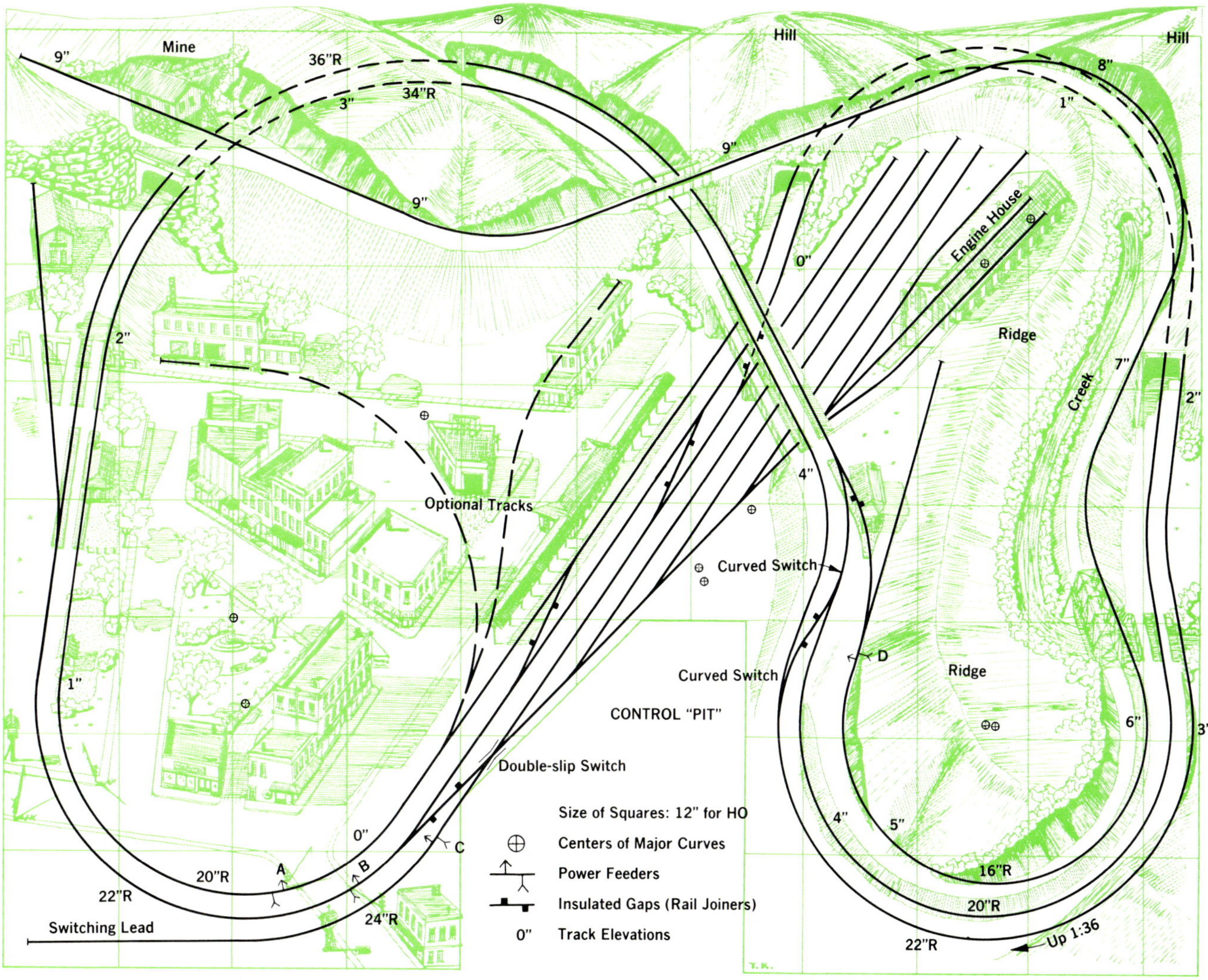

In some instances this is a good thing: the modest length of a model pike is not made to look short by the operation of over-long trains; a short train makes the pike seem longer.

I showed only four electrical control sections, each of which should have had its own power pack. These should have been interconnected *at the track* between all arrow-heads only, or all arrow-tails only. Such interconnections may or may not be needed between two tracks controlled by two sides of a double power pack. Atlas components can also be rigged up with two power packs so that two people can run only two trains anywhere on the railroad (see their booklet on control which naturally is about Atlas products). Of course, each main line track could have been subdivided into sub-sections, but initially at least, I suggested one section for each main, one for the yard, one for the branch.

I tabulated track requirements as follows, subject to changes if you change this design even slightly (few people build any model railroad exactly as planned!), and subject to certain changes if you change to another gauge: 1 #4L switch, 1 #6R, 10 #6L, 1 3-way, 1 double-slipswitch, 2 RH curved; and 129 feet of flexible track (43 3-ft. lengths).

Of course, the layout could be built almost exactly as drawn, using Atlas or Peco #6 switches and so on, in N gauge. That is, make it the same size as for HO, but space double tracks at 1-1/4" or 1-1/2" centers (which would allow you to get in additional yard tracks). The track tabulation would be the same except for increases in quantity of switches and "mileage" of track engendered by added yard tracks.

If the usual proportion is used and the layout is built half the noted HO size for N gauge, the switch count would remain the same, but track footage would be halved as would other linear dimensions: radii, elevations, etc. A big fly in the ointment, in the wheelchair circumstances, would be that the width of the control pit would also be halved, making it difficult if not impossible to enter, depending on your own width. A slim boy could enter, an "old timer" with *his* occupation's version of "Educator's Tummy" could not. Hence, the prospective half-HO version of the Big Eight would really have to be longer to allow for a wider control pit. Along the same lines, 1/2x22"R would be 11" for outer curves, but inner curves would not be 1/2x20"R because N gauge trackage is spaced, where parallel, at more than 1/2x2" centers (HO) or 1". A better figure would be 1-1/4" or 1-1/2" centers. Some sectional N track comes in a nominal 9"R which would O.K. for the inner tracks of curves. Really, to make the entire layout in N gauge sectional track, one would have to redraw it with the range of track parts of the particular brand of N (as well as its compatibility with other brands) in mind. However, a very similar layout could be built, even without a plan, by recreating the general idea of the Big Eight with N gauge sectional components as you go along. The same applies to sectional trackage in all other scales, overall dimensions varying as required.

The Big Eight R.R. could be built in O scale, too, by doubling all linear dimensions, but the eight foot wide lobes at each end would have to be hollowed out with access pits so you could reach track of the expanded layout. 40" radius (2x20") inner curves in O would be a little on the sharp side, but quite suitable for many smaller engines.

In any gauge, a good starting place for the Big Eight's construction would be the double slipswitch area at lower left-center, working diagonally toward the back, right, to complete the yard, and just a little to the left for switching purposes. From back, right, work along the right side, and around the twin curves at the front to the curved crossover. If this comes out slightly different, your big bridge across the yard might come at a different angle. Your escape valve then would be to vary the big 36"R curve at upper left as required to connect the upper end of the bridge with the track rising along the left edge.

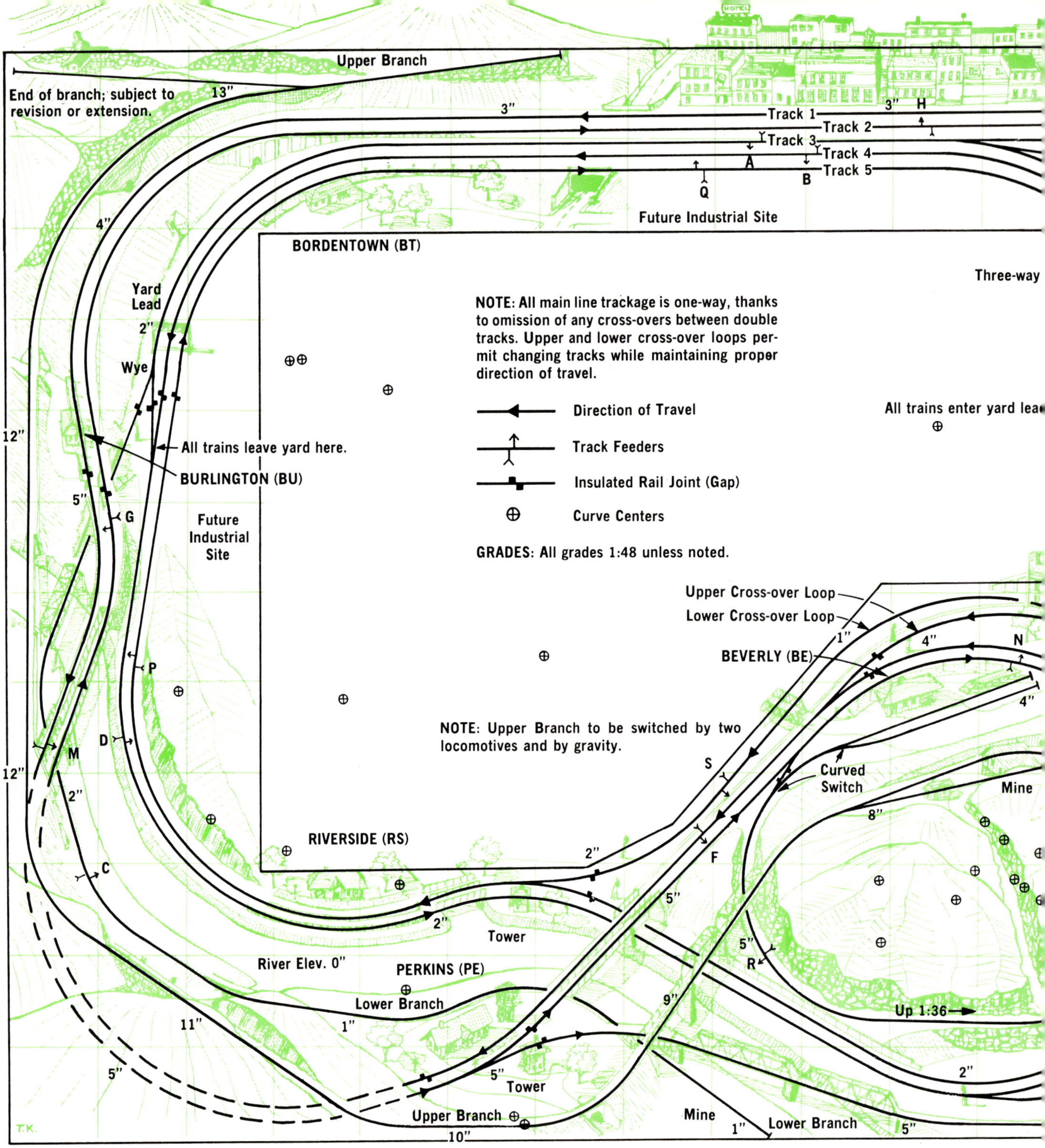

DEVON & NORTHERN RAILROAD

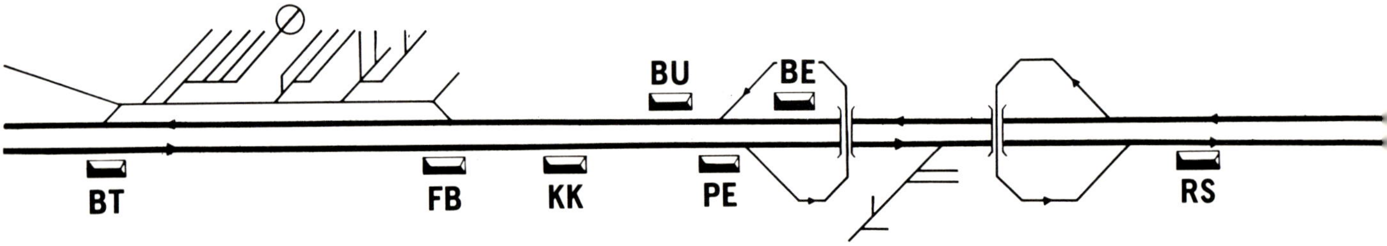

SCHEMATIC PLAN

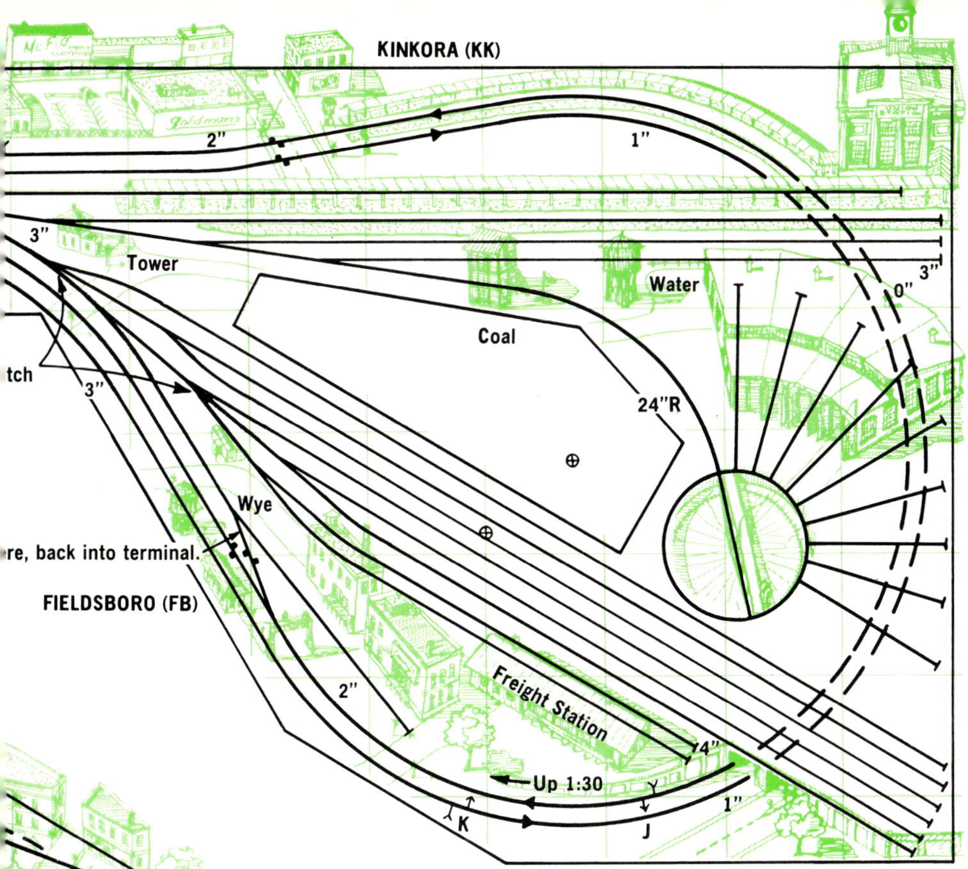

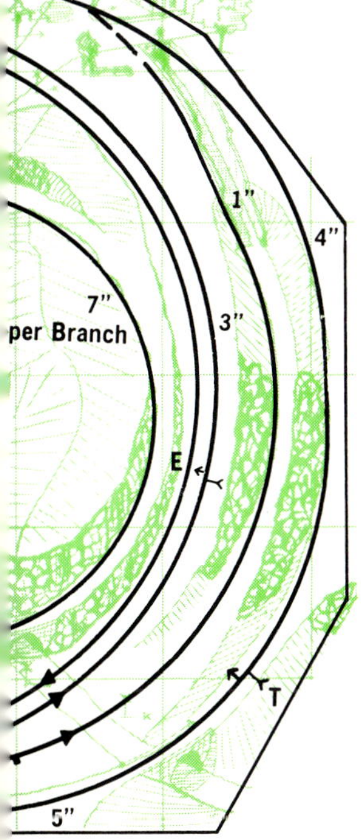

Designed for one-way operation to simply the wiring, the Camden & Amboy could be made two-way by adding crossovers. Scenic possibilities are endless.

The C. & A. line is a large double-tracked HO model pike to fill an area 12' x 20'—double that for O, half that for N. It has curves of ample radii (30" and 28" for HO), gentle switches (#6) and main line grades of less than 3%. It has double tracks, purposely intended for each to have one-way running, thus making automatic signaling (masts and gantries sketched in) quite easy to hook up, whatever system you use. Notice that there are no usual #6 right or left crossovers between the double tracks. There are, however, crossover loops entwined around the mountain at lower right. These, shown diagramatically in the schematic plan, cross your train over to the other main track, and also reverse its direction so that it is running the right way when it gets to that track. Electrically, they need not be considered reversing tracks due to the absence of conventional crossovers.

All trains, freight or passenger, enter the yard area at Fieldsboro crossover to the yard lead at upper right. They can then back into the terminal or freight yard as appropriate, and the engine can go to the roundhouse—which could be revised, without the turntable, for all-diesel operation. After needed switching has been done, all trains leave the lead at extreme left. They can stay on that track, or can switch to the other main line by means of the "lower crossover loop" coming back in at Riverside.

I'd appreciate any comments you may have on my idea, which has cropped up in several plans recently, of having one-way running with the idea of making signaling and ATC more practical. I get a lot of static about it at home where the boys are used to seeing prototype trackage signaled for 2-directional running, but are not concerned, really, with the wiring complications involved in working this out for 2-rail D.C. Do you or do you not like the idea of a pike designed for one-way running?

DIMENSIONS OF LAYOUT:

GAUGE	SIZE OF SQUARES	RADII	SWITCH NOS.
N*	6"	14", 15", 17", 18"	6
HO	12"	28", 30", 34", 36"	6
O	24"	56", 60", 68", 72"	6

*Width of aisle between Fieldsboro and Beverly must be widened for N scale.

CONSTRUCTION SEQUENCE:

STAGE	SWITCHES	FEET (FOR HO)
I (Yard and Lead, Terminal and L.B.)	4R, 4L, 2Y, 2 3-way	120
II (RS to KK via BT)	1R, 1L	72
III (KK to RS via PE)	2R, 2L	100
IV (Upper Branch)	2R, 2R-curved	110
TOTAL	9R, 7L, 2Y, 2 3-way, 2R-curved, 402 feet.	

by the layout doctor

A Railroad Designed For One-Way Operation

Point to Point

by the layout doctor

Though the basic design and scheme of operations call for point-to-point utilization, an "interchange" track provides a loop for continuous running and loco break-in.

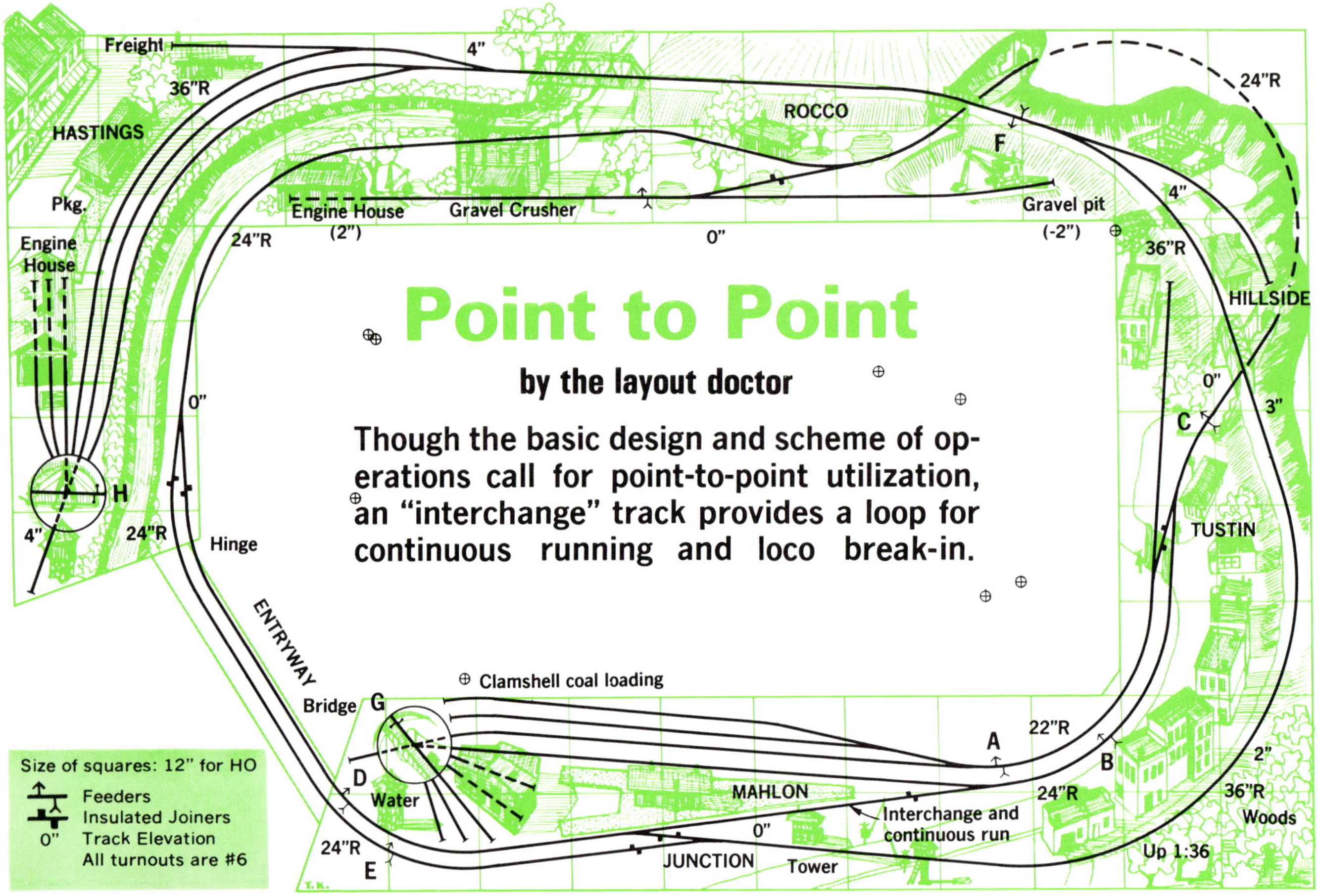

MAHLON & HASTINGS RAILROAD

Given a room 9'x12' for an HO pike, the average model railroader (myself included!) would build a thriving Class I railroad with as much trackage jammed in the area as possible. There are other fellows who are more enamored of the single track short-line railroad which does not fill every square foot of tablework with track. These fellows prefer a one-(iron)-horse railroad, dying in the Thirties, or even still holding on today thanks to careful management and benevolent ownership. No 4-8-4's or 2-10-4's ever ride the rails of such a pike—they'd turn over, or crack the rails. Nothing with heavier axle loading than a Russian Decapod, a light Mikado, a Mogul or Prairie will be seen; or, if it is a present day line, perhaps a 44 or 70 ton GE diesel switcher (maybe even a BL-2 or A1A-A1A Baldwin), with steam reserved for heavy movements and fan trips.

The Mahlon & Hastings is such a small time railroad. To make it look even more so, I'd suggest using code 70 rail spiked to individual wooden ties set into dirt ballast (certainly not clean gravel!), with all-rail frog switches. It should have plenty of weeds between and around the rails to give that appearance of desuetude common to rail lines which run only once or twice a week (although the model railroad would run more often).

One of the aims in designing the M&H was to keep the number of switches low; I think I have succeeded—seven each right and left-hand #6's (and 111 feet of track). One "trick" in doing this is to use turntables at the ends of the line, letting their turning accomplish switching. (At each end I also show dotted alternate arrangements not serving engine houses, for switching engines around trains). These turntables need not be automatic, for a real one-horse railroad would have "armstrong" turntables; a reproduction of such a line would have "fingerstrong" tables. I figure them to be 9" or 10" in diameter, and at least one would be the gallows type like Cliff Line makes in HO. This will make things much less complicated. Operators at both ends of the line could each manage his own, an extra operator in the entryway could be assigned to turntable turning, or a lone operator on a well castered swivel chair could zoom back and forth in the central operating pit and work both.

Each turntable has a radiating enginehouse that will hold more locos than such a short-line would really have. This is because model rails eventually acquire more engines that they really have need for, and should be able to store them undercover. Both ends have ash-pits and water cranes served by adjacent water tanks. Coaling at both ends is done by clamshell cranes on short stub tracks lifting coal from gondolas into the tenders (this after original coaling towers had to be demolished as unsafe).

Starting from Mahlon, where there is a track connection with "another railroad," the line runs past Tustin, then through a tunnel to Rocco, named for the Rocco Gravel Co. Thence along the low bank of the creek, crossing the entryway on a double track lift bridge. This is the passing siding needed half-way along a single track point to point line, although I guess if only one engine is "in steam" at a time, it could be reduced to single track with evidence of double track still showing. At Junction there is again a connection with "another railroad" at Mahlon. This is to permit continuous running on the level oval just described, as well as interchange of cars from one railroad to another, though they are really the same pike.

From Junction the single track proceeds up a 3% grade to the right. A heavier loco or even a geared engine might be used for the hill-climb, though it really isn't bad as model railroad grades go. At the top of the grade there is a station cleverly named Hillside, followed by a high level run across a road and river to Hastings.

The Rocco Gravel Co. has some interesting possibilities. At first I pictured it as a level (0") line operating its own lone steam engine to bring chunks of rock from the gravel pit to the crusher, and then help take these to main line trains. However, by keeping it in a straight line and using the elevations in parentheses [(2") to (-2")] the line could be operated by a cable, winched up by a drum inside the "engine house" which would be a *stationary* engine in this case. If the uniform table width of 2' were widened only 6" for the top table, there would be additional space to expand the Rocco Gravel Co. with more buildings and tracks, perhaps even some HOn2.

It is possible then that the industry at Tustin could be included, maybe with a line running along the inner table edge into the empty area at Mahlon.

Walk Around Cab Control

For many modelers, walk around cab control provides the ultimate in operation once the railroad is completed.

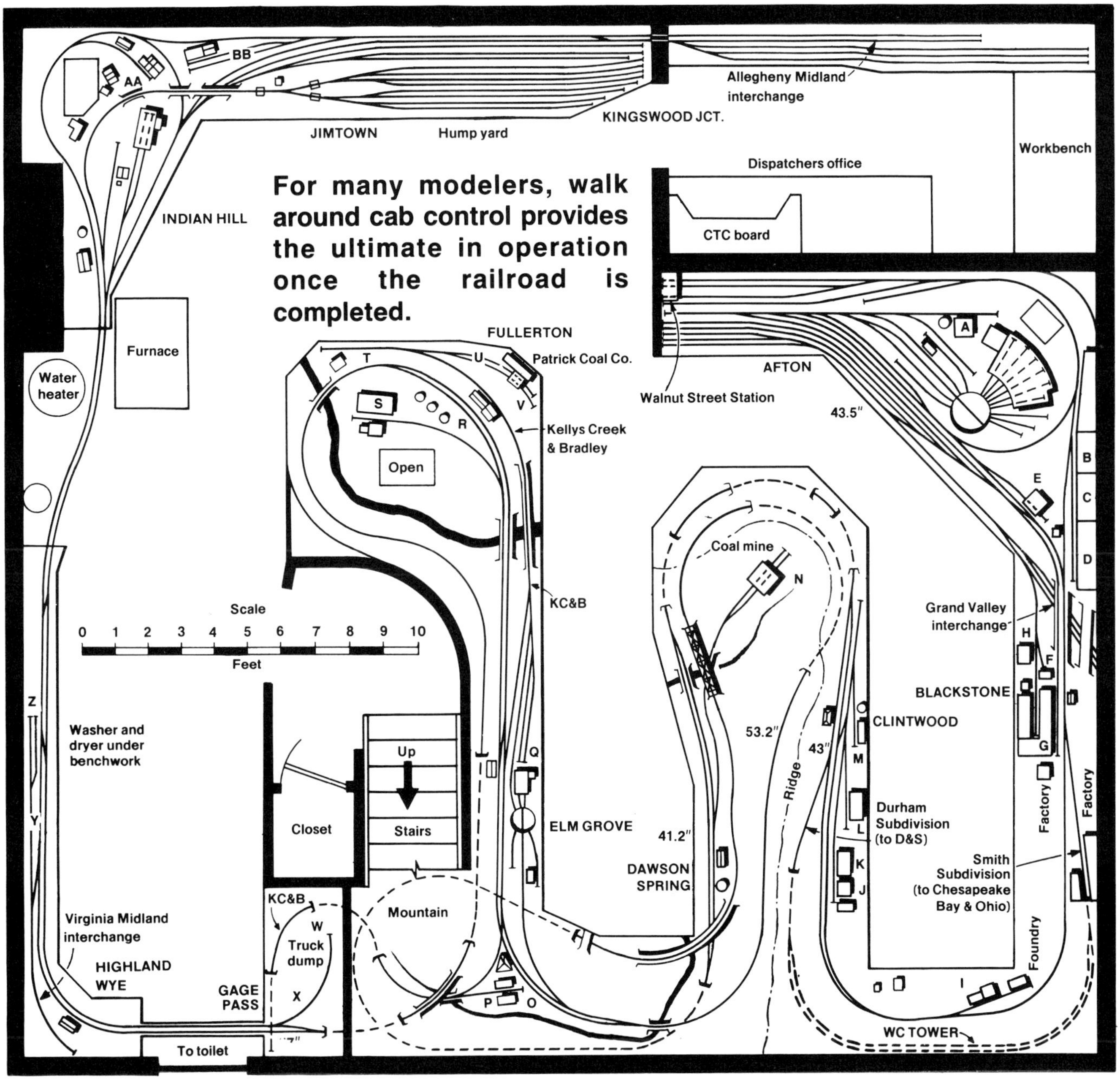

Allan McClelland's Virginia & Ohio Railroad is an HO railroad which was completed years ago but has never been finished. It is always changing and improving. Periodically Allan updates the time era in which the V&O operates, requiring substantial changes not only in the locomotives and cars but in the railroad and other structures found on the railroad itself.

The V&O was designed specifically for walk around controls, in which the operator or engineer walks around the layout with the train at all times. Control of the train rests in a hand held controller plugged into outlets mounted along the aisles in the train tablework. The operator can start or stop the train, back it up, and control the speed. He can do switching when and if necessary. Each locomotive on the V&O has a different frequency allowing operation of many trains at any given time, and providing superb opportunities for time table train operation which many advanced modelers claim is the ultimate thrill of model railroading.

Notice how the design of the railroad utilizes narrow shelf work along the walls and relatively narrow layout peninsulas broken up with lots of aisle space. The result provides a long point to point run from Afton to Jimtown or the train can be reversed at Indian Hill and return to its starting point. The KC&B is a branch line connecting railroad having its own engine house.

On the track plan, the letters A through BB indicate the many factories and railroad freight houses which provide revenue for the V&O through car movements both in and out. Interchange facilities indicate points at which other model railroads connect with the V&O providing interchange of freight cars. Located in the McClelland basement, the V&O is subject to the same space and obstacle problems faced by most modelers. The complete history of the Virginia & Ohio with many full color photographs telling how Allan developed the original concept, built the railroad, and how he operates the railroad with his friends is detailed in his book The V&O Story (C47, Carstens Hobby Books) available at most hobby shops.

Railroad For a Spare Room

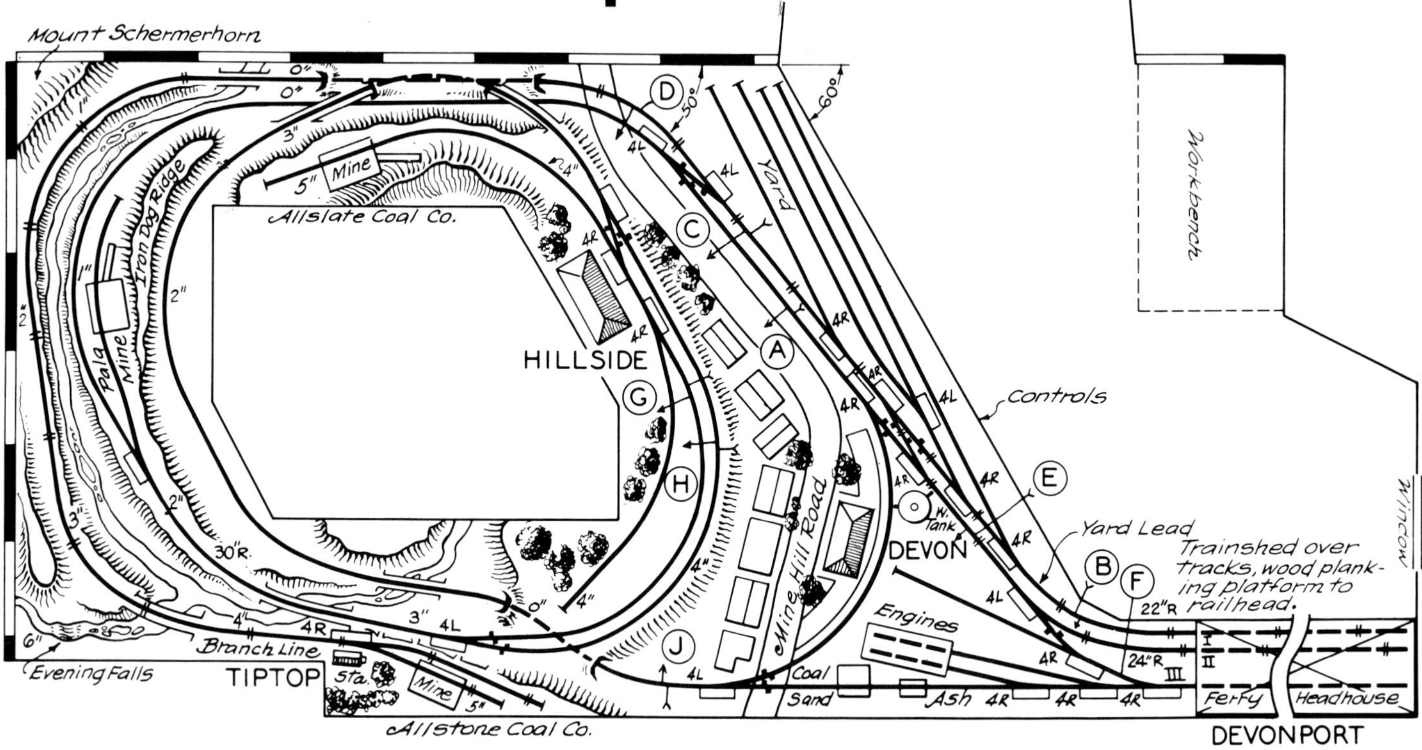

Drawn for RMC by James Berge

Designed to fit a spare room, this layout provides ample entry room and access to a window while still leaving room for a workbench.

In planning a layout for a "rectangular" room, it is often a good idea to allow full and free access from the door to the window, as shown here for this room, which measures (less door alcove) 6'9" x 14'0" over-all. Such a free passageway not only gives a good site for a workbench or cabinets, but also an "outside" control position, and an area for a few spectators. The alternative would be to fill the main part of the room with an around-the-walls shelf (18" to 24" wide) with the door and window alcoves free for access to them, as well as to the outside of the railroad. The center then would be the main control position and everyone would have to duck under to get in and out.

By utilizing the free passageway idea, one side of the window alcove (which could be a dormer) can be made part of the layout by running a terminal up into it on a narrowish board. This is really where I began this design, running in three terminal tracks to a ferry headhouse, steamboat landing, or a small city terminal building. All three tracks are floored with planks to railhead height, and this forms all the passenger "platforms". The terminal area—tracks, platform, headhouse, and trainshed—are shown slashed across because it would be very nice to have them longer if space permitted.

Two tracks swing out of the terminal toward the top of the plan, one becoming the single track Devon & Northern main line, the other (also acting as yard lead) becoming the branch line. The third terminal track runs straight along the bottom wall and enters the main line at center for running the other way. Thus, with suitable switches just outside the trainshed, we have a wye, the use of which will be mentioned soon.

The main line itself is nothing but the ol' inside-out figure-8 or twice-around oval, with a single passing track at the top of the grade at Hillside. Curves are 22" and 24" radius, and grades are 3%—1" rise in 36"—or slightly less. Clearances where one part of the main passes over another are stated at 3" difference in elevation, but I suggest that you use very thin baseboard where clearance is required—1/8" Masonite is good, thick cardboard (or even thin cardboard) with the rails supporting themselves for the few inches just above the underneath track. Track required for the main line is: 5 #4R switches, 3 #4L, and 47' of flexible track.

(Track required for yard, terminal, and branch: 9 #4R, 3 #4L, and 49' flexible track. Total track required for railroad, 14 #4R, 6 #4L, and 96' flexible track.)

Terminal track I is intended for inbound and outbound branch line trains which, when no switching is going on even though the main line be in use, can run independently. Terminal track II is intended for outbound trains headed out on the main counter-clockwise. Track III is intended for trains outbound in a clockwise direction on the main. Counter-clockwise trains arrive on III, clockwise on II. Then, if they have locomotives and need reversing, they are backed out to the main, around the wye and into their designated outbound tracks. When there is need for switching on the yard lead (Terminal track I), trains which normally use it will use track II instead.

Arrival and departure of freight trains will be more complicated because of the location of the RH crossover below the wye switch, but its very complexity will lend operating interest. Naturally a second (switching) engine, besides the road engine will be needed, and the wye terminal track III, will enter into the picture.

The would-be builder of this Devon & Northern also owned some interurban equipment and wanted a place to run in on his otherwise steam/diesel line. My suggestion is to electrify the branch line as well as terminal track I—and II for when I is blocked. Of course a real line would not permit overhead wire above a yard lead, but since it's only 12 v. instead of 11,000 v. and the people are plastic instead of flesh & blood, perhaps it will be permitted here. What I did point out to him is that once you allow overhead wire to intrude on your layout, and get it running smoothly, it's not long before you string overhead over the former steam/diesel tracks. It's the old story of the camel getting his nose in the tent!

HO – 10½′ x 11′

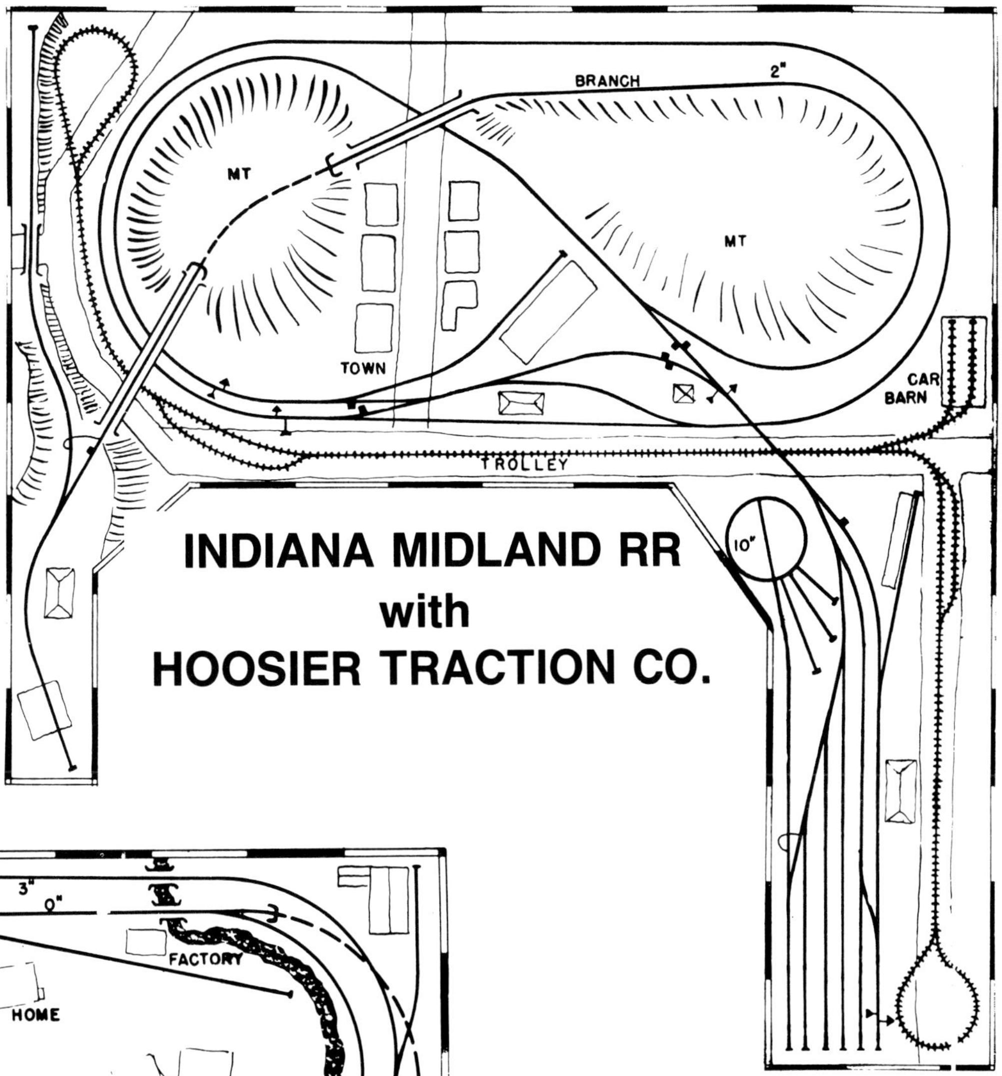

INDIANA MIDLAND RR with HOOSIER TRACTION CO.

The moderately sized Indiana Midland offers varied operating possibilities including point to point, out and back, and continuous. While not shown, there could be a freight interchange with the trolley line.

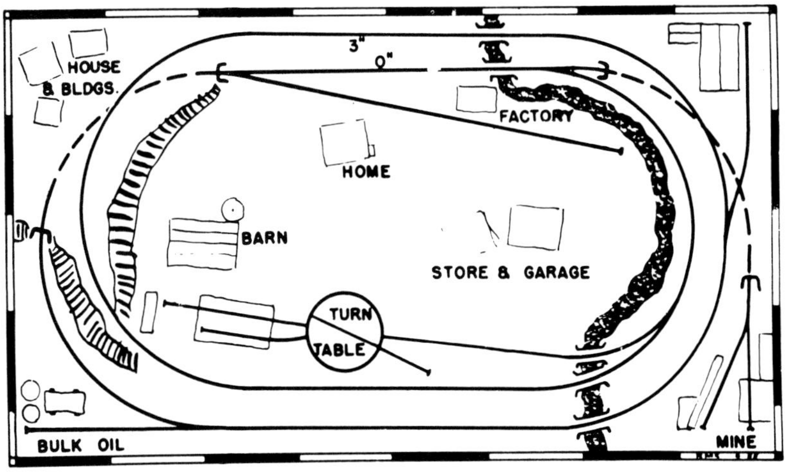

CALIFORNIA RR - HO

Two levels of operation are provided on this beginners pike which is easily extended at either end. There is ample room for more sidings and industry

DIXIE CENTRAL RAILROAD - O

The Dixie Central was designed as a small O scale pike but adapts to larger and smaller sizes, packing in a turntable, yards, industries and a mine. Open space is deleted in smaller scales.

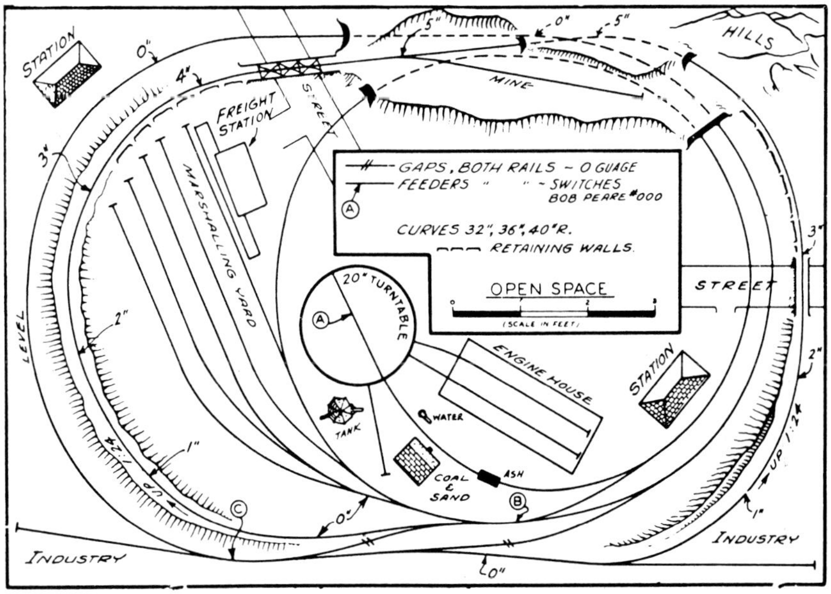

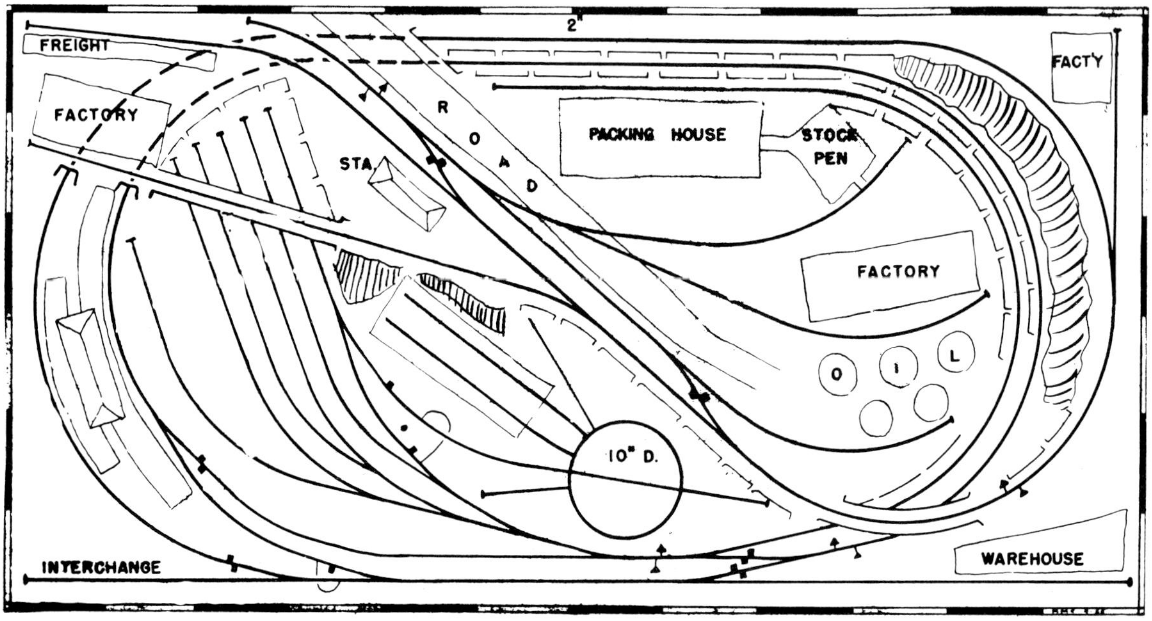

LINDEN INDUSTRIAL RAILROAD - HO

Plenty of railroading is packed into the 4x8′ Linden Industrial Railroad. Passenger service on a pike like this would me meager at best, perhaps a gas-electric or a few shorty coaches hauled by a small diesel switcher. Scenery can be changed to suit your needs.

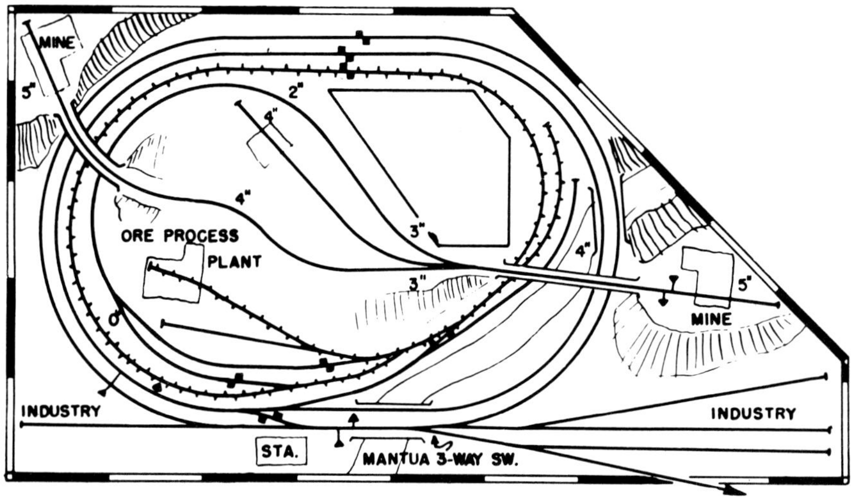

BOOTHWYN COAL RAILROAD - HO

The Boothwyn Coal Railroad demonstrates how any oddly shaped space can be converted into layout space. This one has a double track main line, branch line to the mine, and a narrow gauge feeder.

LEADVILLE & TORRANCE RR - HO

The Leadville & Torrance is a narrow gauge railroad bringing ore down to the standard gauge railroad connection at the bottom of the layout. Think about using the same design and space for N scale. Use the same space for On2.

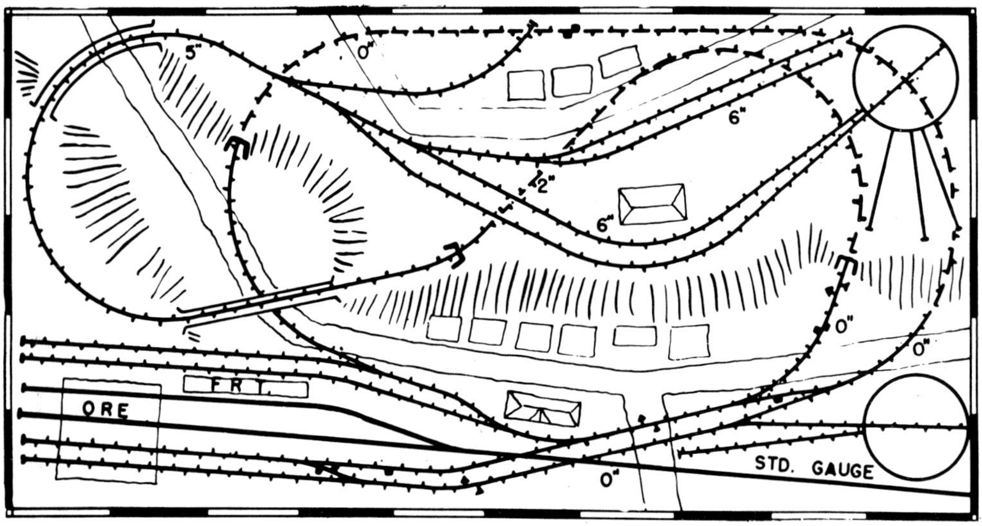

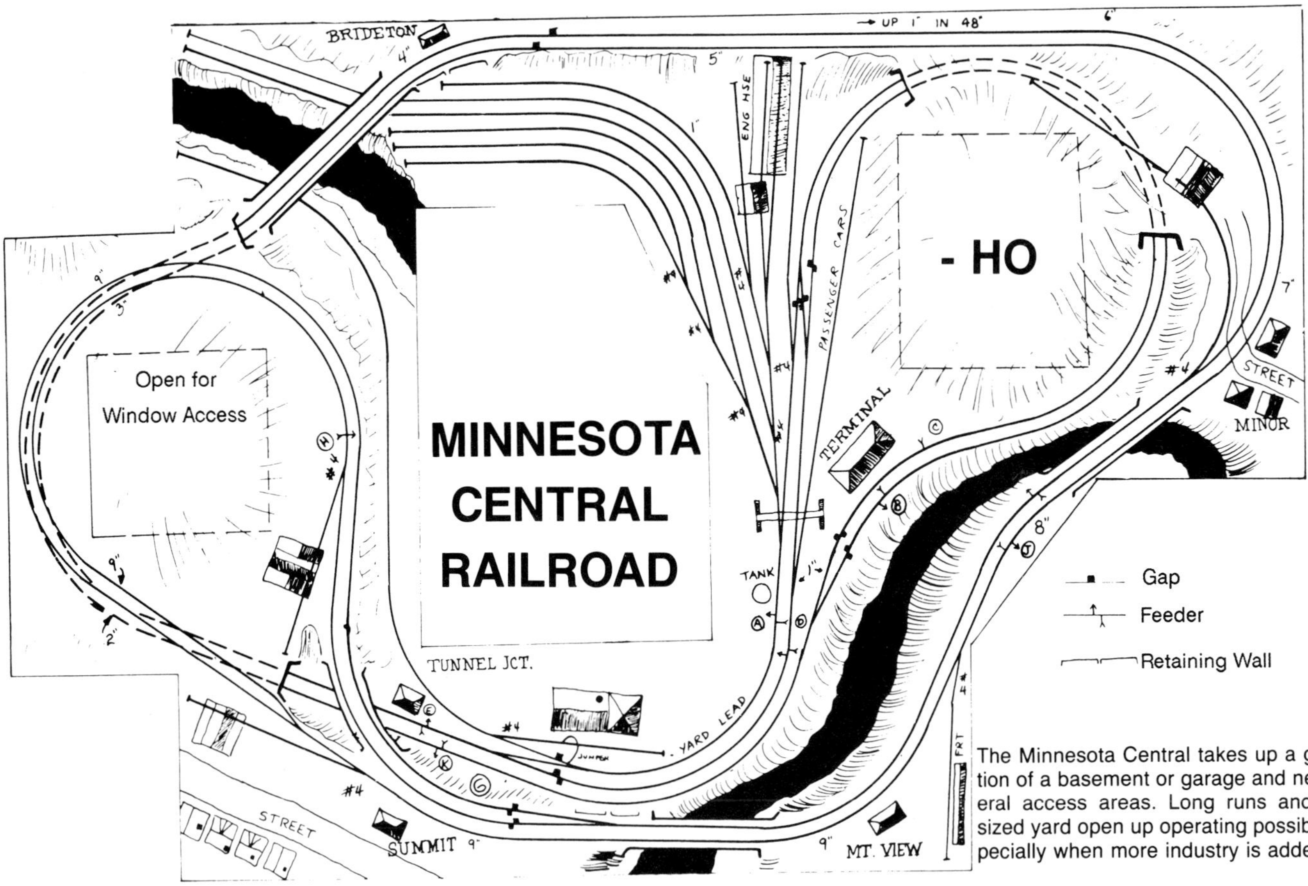

The Minnesota Central takes up a good portion of a basement or garage and needs several access areas. Long runs and a good sized yard open up operating possibilities especially when more industry is added.

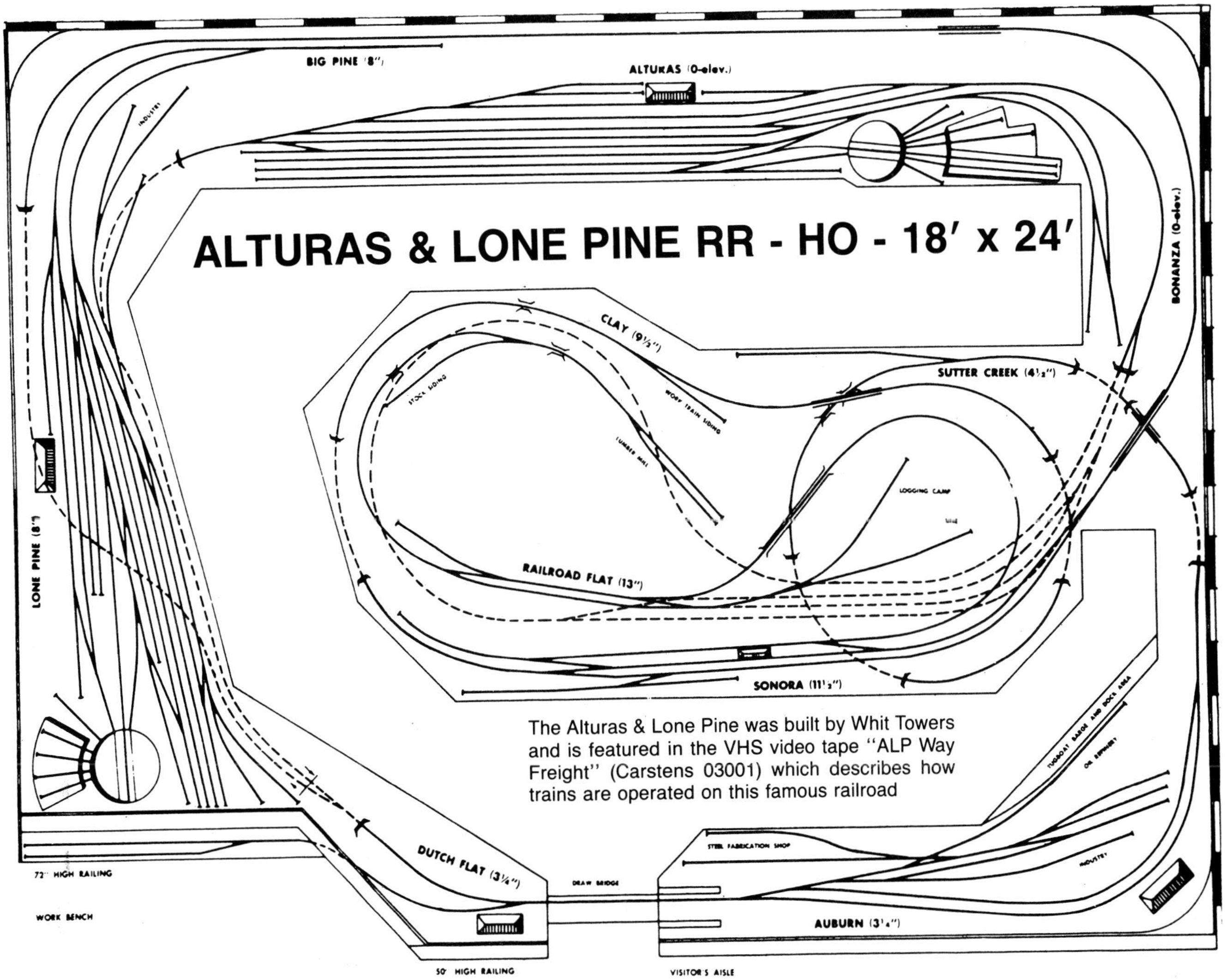

The Alturas & Lone Pine was built by Whit Towers and is featured in the VHS video tape "ALP Way Freight" (Carstens 03001) which describes how trains are operated on this famous railroad

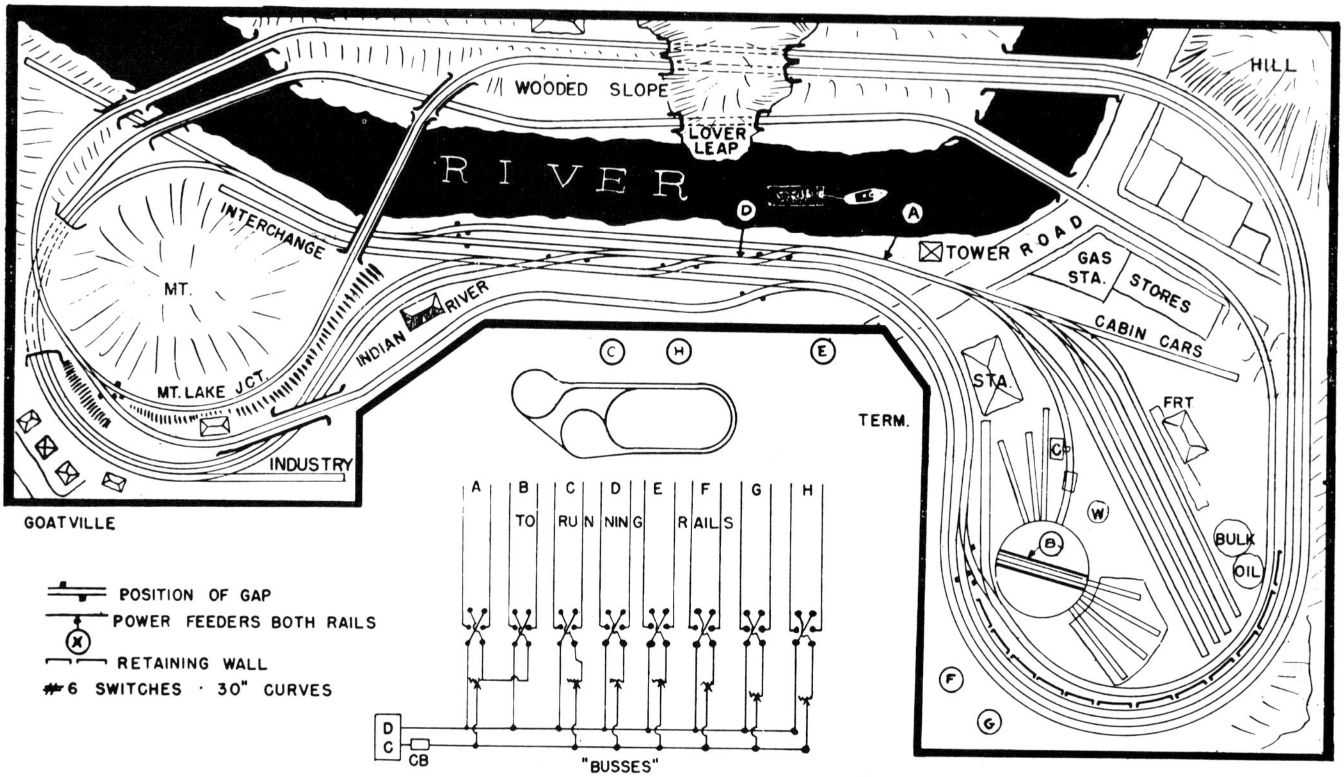

OHIO RAILROAD - HO

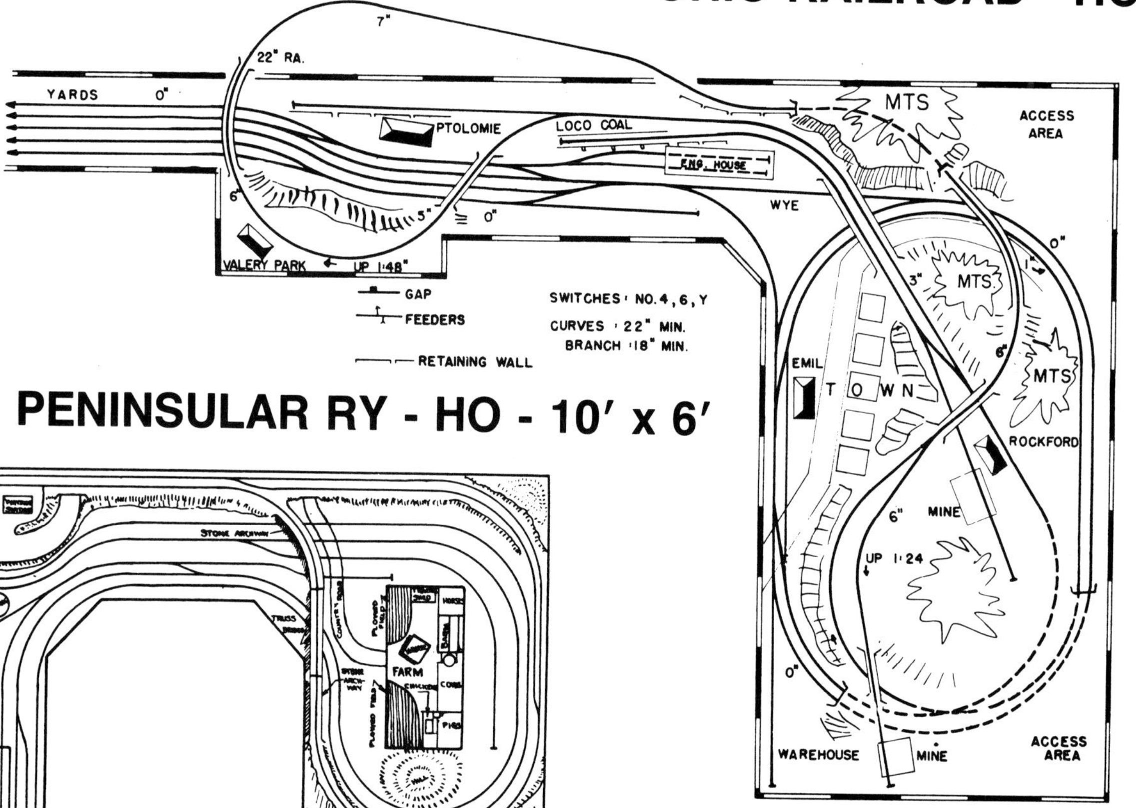

PENINSULAR RY - HO - 10′ x 6′

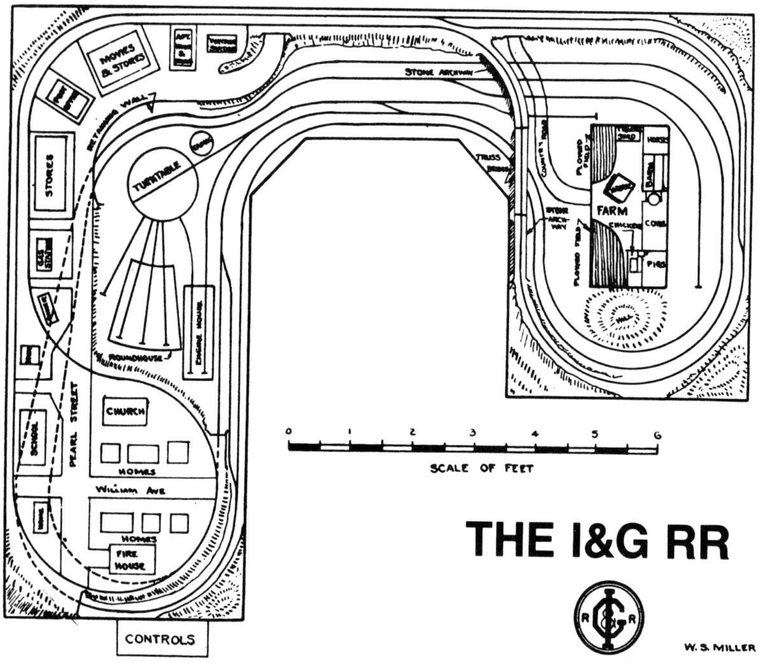

THE I&G RR

The Ohio Railroad at top is divided by a river making mandatory a number of bridges. The Peninsula Railroad, center, can have the yards extended to the left depending on your available space. Notice how the loop at Valerie Park goes behind the wall, sometimes possible when the space behind the layout is a laundry or deadspace. The I&G RR, bottom, has several points where the railroad could be made larger by extending the tablework.

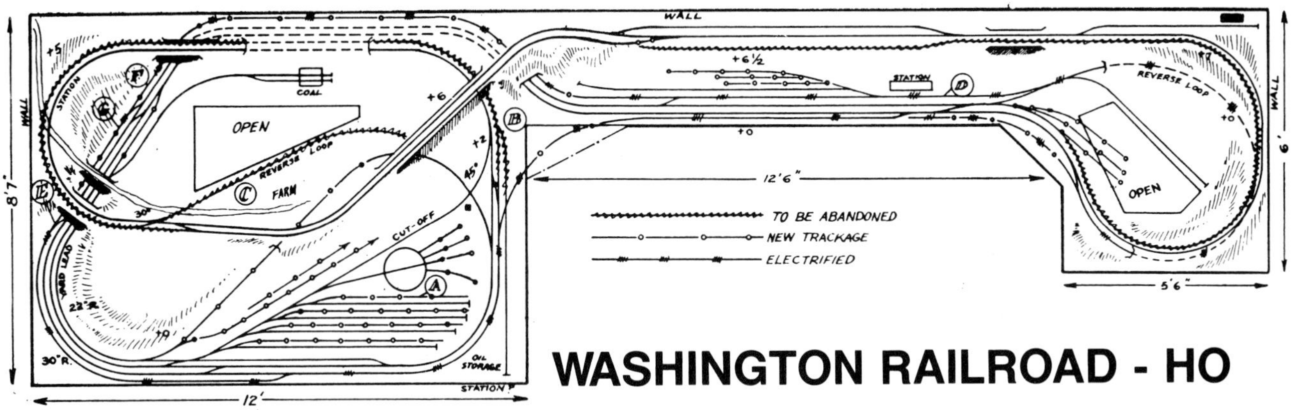

WASHINGTON RAILROAD - HO - 30′ x 9′

Ample yard space in these three track designs provide space for storage of cars or complete trains. The Washington Railroad at top shows again how a moderately sized layout is extended to provide longer runs on narrow shelf space.

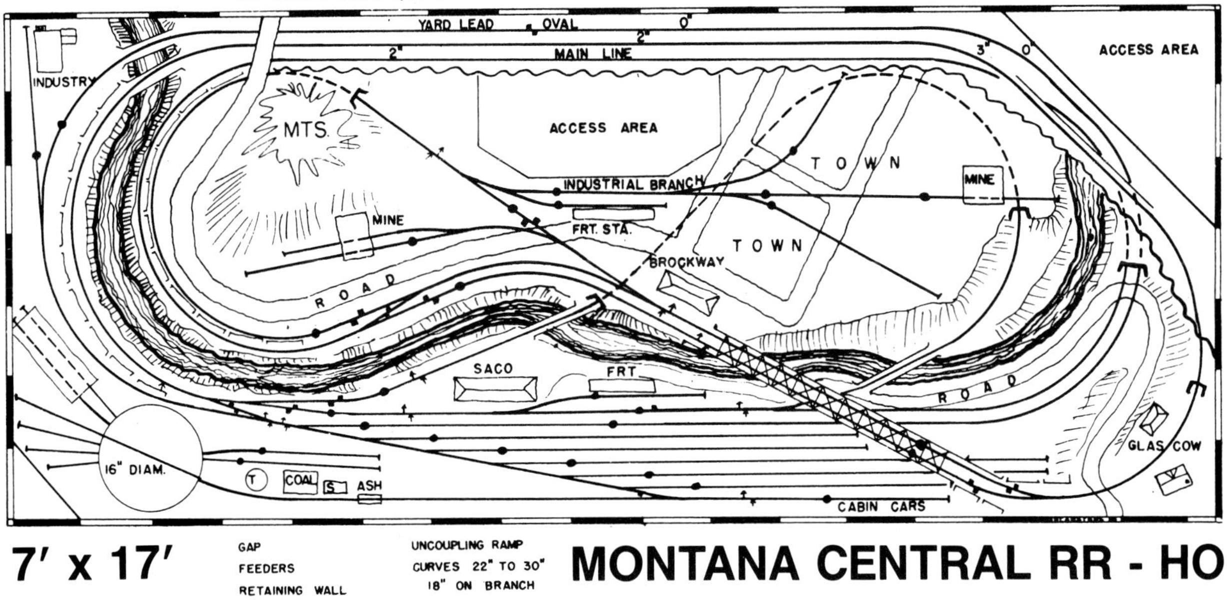

7′ x 17′ MONTANA CENTRAL RR - HO

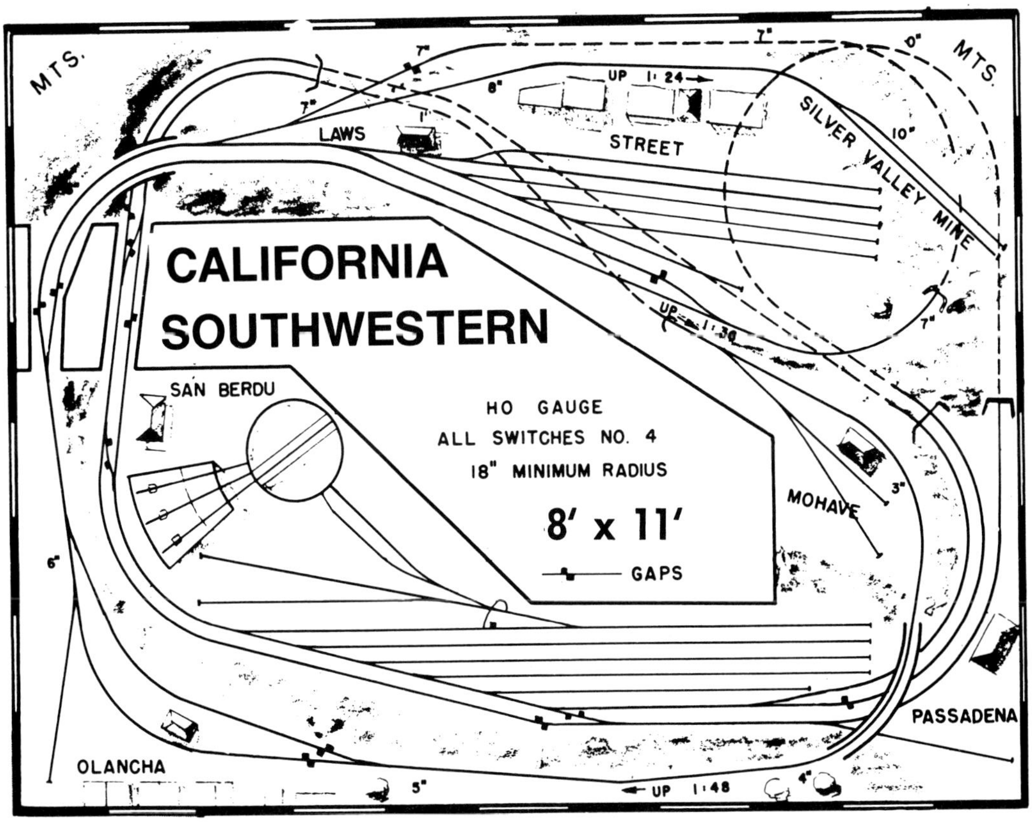

CALIFORNIA SOUTHWESTERN

8′ x 11′

Reversing Sections

by the layout doctor

In the early days of model railroading, nearly everybody ran his trains from third rail—inside or outside—or overhead wire. 'Twas said that two-rail was too complicated when it came to insulating engines, cars and switches—and in wiring reversing tracks. In spite of such warnings, there *were* two-rail pikes, such as Harry Albrecht's O gauge PRR in the mid-'thirties, not to mention some of the early "tinplate" lines.

Model rails are an inventive group and just before the War more and more attention was given to two-rail. Even with one electric loco to "justify" the outside third rail on a "steam" railroad, many modelers did not want the third rail shoe sticking out from every engine. Insulated metal car and driving wheels became available for both the Varney and Tyco range of locos at, originally, slight extra cost. The system of two-rail in wide use today, with all-rail frog switches, was expounded by Jim Dechert and Bob LeMassena, helped along by Eric LaNal and others.

But the problem of reversing sections of track—reverse loops and wyes—remained. Not much of a problem, to be sure, but one which needs elucidation now and then for the benefit of newcomers to the hobby, as well as those old hands who claim to be all-thumbs electrically.

A reverse loop sends an engine, in continuous motion, back over the track from which it came. In Fig. 1, an engine from the left (the main or "stem" of the loop) would round the reverse loop and go back whence it came. The switch would have to be turned, although if it were a trolley car, a spring switch could be arranged. Disregard the four black squares which indicate insulated rail joiners and disregard the DPDT and feeders to rails at right; pay attention to the polarity symbols (plus and minus) in parentheses. A train coming from left would pass the switch and round the curve at right, slant diagonally toward the left again; the switch would then be thrown and back it would go. Or would it? For with polarity of track as indicated, the plus on the bottom rail would (as it swung around hit the minus of the top rail) short out. Nothing would run. There would be a short circuit and the breaker or fuse would blow on the power pack. About all one could do would be push the engine past the switch.

Because of the impasse above, the reverse loop must be isolated with the four (two pairs) insulated rail joiners, with the power fed separately. When the polarity of the reverse loop is lined up with the power on the approaching main line, the train should proceed from one to the other smoothly. Of course, the DPDT direction switch should be set the same—one has a 50-50 chance of being right even blindfolded, and the proper setting is quickly mastered. Even an error should result in little harm, except a stalled or shuttling train. The rheostat must be set at the same "speed" as the main line.

Picture the train *on* the reverse loop. It's going to have to emerge, so the turnout is thrown. At the same time, the polarity of the main line is reversed, indicated by non-parenthetical plus and minus signs. The train should proceed right out onto the main.

It is helpful to have the main line and the reversing loop fed power by the same D.C. power-supply, which might mean a little wiring. If they have different power supplies, no great harm will be done, but it is possible that due to poor contact the polarities maybe correct, yet the train will stall, since power packs (like separate batteries) are completely independent. If separate power packs are used, then instead of having the pair of insulated rail joiners approximately opposite each other, stagger them with the right-hand one in advance of the left one by about the pick-up wheelbase of the average locomotive.

Fig. 2 shows a wye. I have not made little plusses and minuses here, but it can be easily seen that, say, the "bottom" rail across the bottom track, through switching back, then forward again through the wye, would effectively become the "top" rail across that bottom track. This illustrates the need for insulated rail joiners as drawn and the separate reverse switches (DPDT) and rheostats for both the main track through the wye as well as the stem.

The devil of it all is reversing loops and wyes are seldom as simple as just described. I see a lot of prospective and already built layout plans and sketches which won't run because the owner has failed to detect a reversing section. Often he has far too many of them, a matter of personal choice perhaps, but quite an unnecessary nuisance in wiring and buying controls. If one starts tracing with a pencil all around a layout plan, making little directional arrows by the track every few inches, he can discover any hidden reversing sections: just note any location where in going forward or backing across to another track, there are arrows pointing in the opposite direction. This indicates there's a reverse track somewhere. Merely "backtrack" to locate it.

After each reversing section (Y or loop) is located on a trackplan, there's no real mystery when it comes to wiring it for smooth operation.

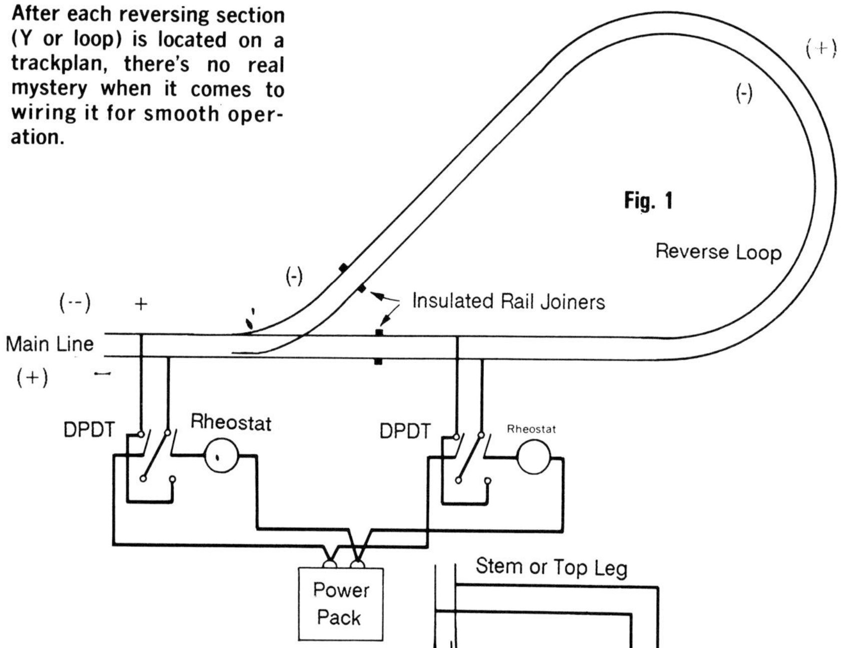

Fig. 1

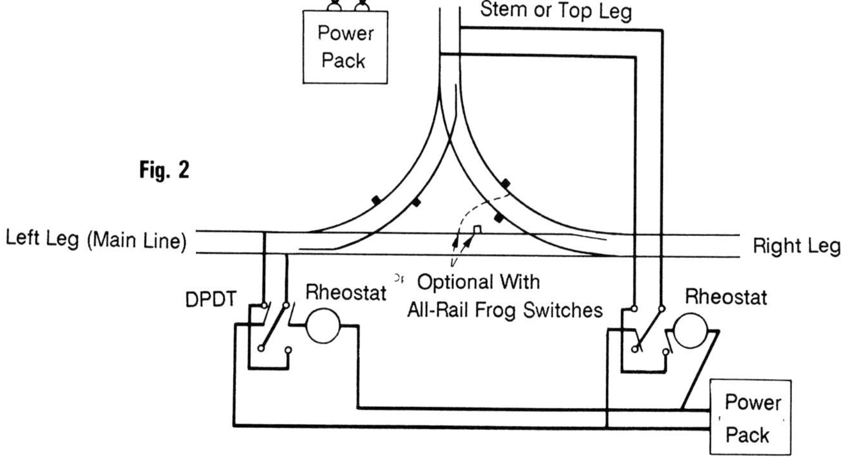

Fig. 2

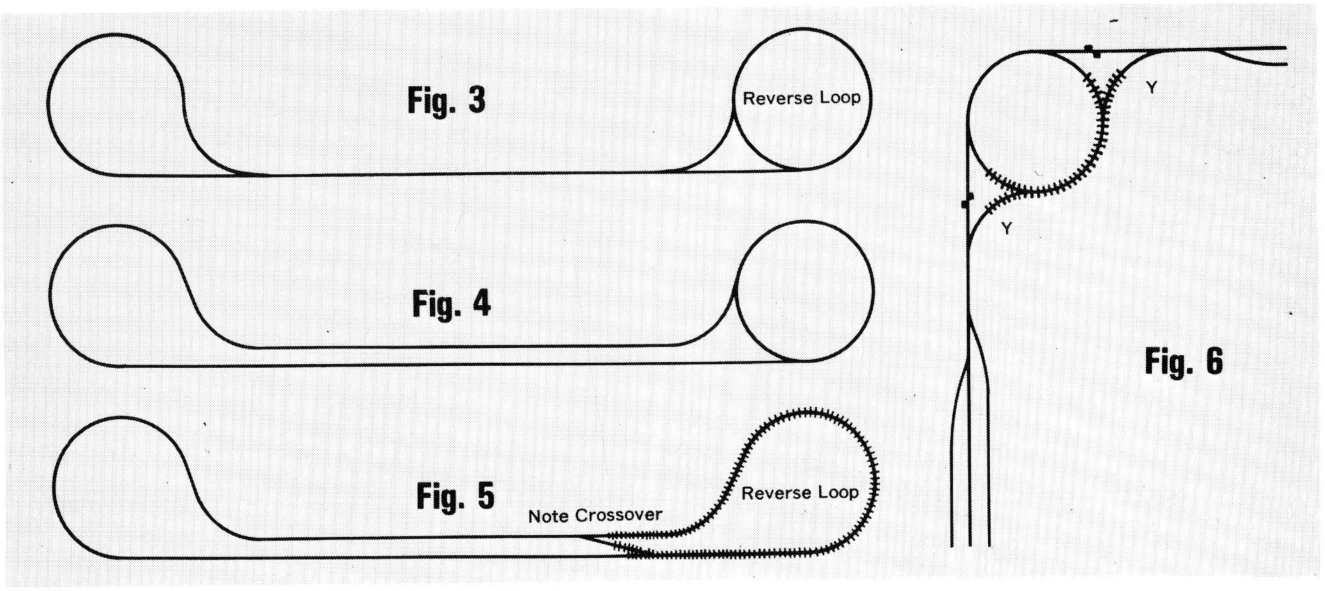

Normally, I find reversing sections by noting the direction of the arrow heads and tails indicating track feeders. I start out with an arrow head pointing to the inside of the layout and, when I get to the other end of the circuit, the same should still be true, arrow pointing inward. But on one layout I'd studied quite a bit because it especially appealed to me, this wouldn't come out right in spite of checking and re-checking. There was a reversing section all right, cunningly disguised, which I finally discovered.

Fig. 3 shows a sort of classic reverse loop situation, a long single track down one side of the layout room, with a reverse loop at one or both ends. Passing tracks, sidings or branches will not change it. Here both reversing loops as well as the lobes at each end must have separate control.

Fig. 4 shows a similar layout, but double track. It does not have reversing loops, electrically, for there is no connection between the double tracks. If, however, as in Fig. 5, a crossover is placed between the double tracks, there *is* a reversing loop (here, really two blocks which must be separately controlled).

Fig. 6 shows how wyes and reversing loops can be combined in what may seem a puzzling way. Actually, only the cross-hatched track need be regarded as a reversing track, as well as a combined stem for both wyes. I show insulated rail joiners on the straight legs of both wyes, because they will probably prove quite handy for more independent control of trains. On this "pike," there would be four control sections, including the reversing track.

Using commercial power packs and such, there is hardly a way one can get in real trouble with a reversing track. At the worst, a train will not run (which may be due to poor contact or any of many other reasons). Even if a wrong connection is made, the circuit breaker or fuse on the power pack will blow before expensive damage is done.

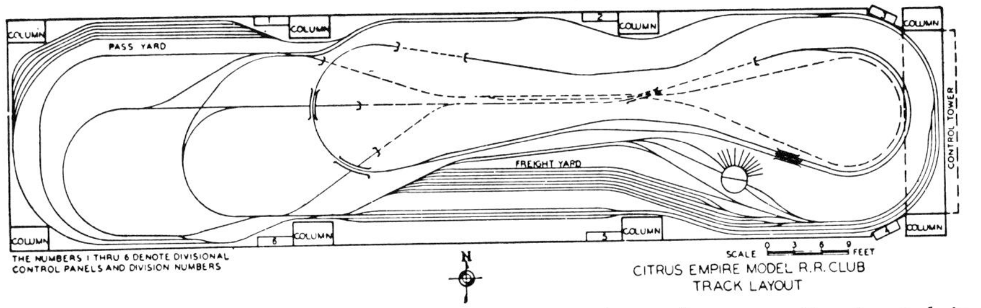

CITRUS EMPIRE MODEL RAILROAD O Scale

Track layout of the Citrus Empire Model Railroad, an O gauge pike located in Pomona, Calif.

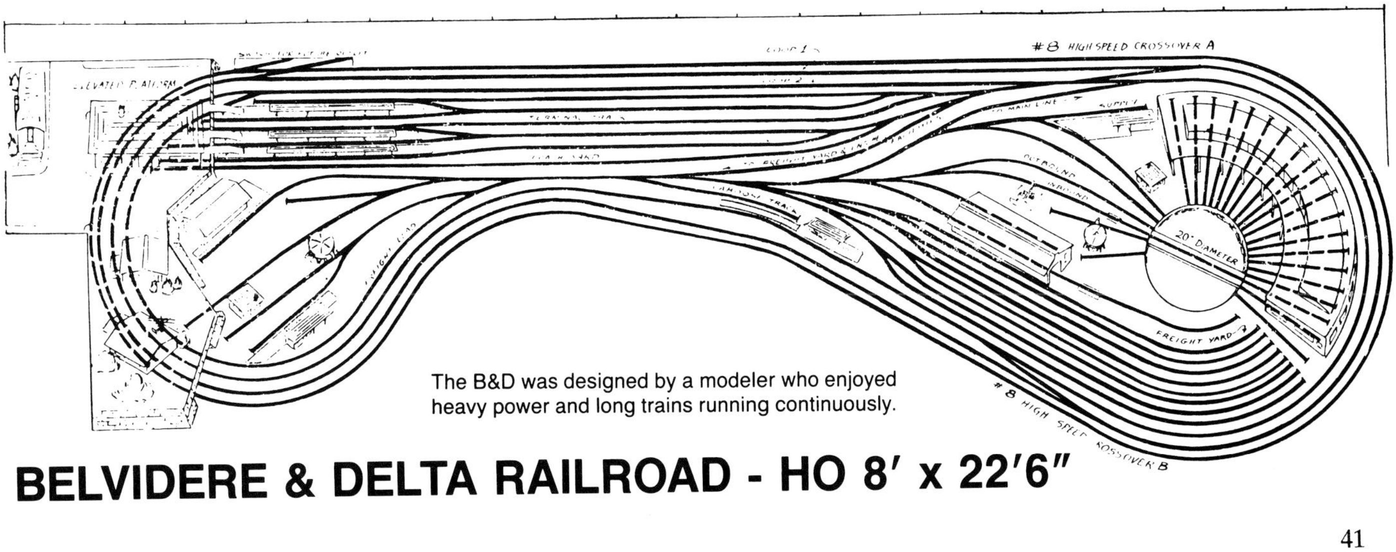

The B&D was designed by a modeler who enjoyed heavy power and long trains running continuously.

BELVIDERE & DELTA RAILROAD - HO 8′ x 22′6″

Eye Level Layouts

by Don Fiehmann

Most of us have seen or owned a model railroad at one time or another that fit the description of being a "Bowl of Spaghetti." Have you ever stopped and tried to analyze why the layout looked that way? Wasn't it a layout you looked *down* on, a pike where you could see the tops of rails running every which way like spaghetti in a bowl?

Have you ever looked at a railroad scene, or maybe a yard, from an elevated spot like a tall building, hill or low flying plane and had the feeling that you were looking at a model instead of the real thing? If you have had this experience, remember that you were *looking down* once again. Recently, three members of "The Nites of The Roundhouse" (a model railroad club) chartered a small, high-winged airplane and took pictures of the Southern Pacific "Del Monte" and "Daylight" passenger trains from the air. Many of these pictures looked like model shots and not like the prototype. How often have you seen an aerial photo of a railroad in a magazine that made you look twice to determine whether or not it was a model?

Most layouts are built somewhere between 36 and 42 inches off the floor. Most people are built with their eyes between 60 and 70 inches off the floor. That means the difference between the average layout and eye level is about two feet. (Have you ever noticed that most people seldom sit when viewing a layout?) This two foot difference ends up to be about 90 feet in O scale, 174 feet in HO scale and 320 feet in N scale. That difference is about the same as the elevated view point you get from some tall buildings, hills, etc. But it's not normal to look *down* on things from an elevated place. What *is* normal, when looking at the real thing, is to stand with your feet on the ground and look straight ahead or, in many cases, to look *up*. From this vantage point you see the sides of buildings, the sides and underbody details on rolling stock, and all of that nice action on steam locomotives.

Have you ever leaned, or sat, so your nose was on the top of a layout? If you haven't, you should, because when you do, suddenly a whole new world opens up before your eyes. The model then begins to give the illusion of being the "real McCoy." Have you noticed that the best model pictures are usually taken from a low angle? I once saw an 8mm movie taken from an O scale flatcar while it was going around a layout. The parts where the camera went past a passenger train in the siding and thru a truss bridge were fantastically realistic. You could see that it was a model, but it gave you the feeling of riding in an open observation car and watching the rails go by.

My first layout was a 5 ft. by 9 ft., and 40 inches off the floor. It looked like most of the layouts that could be classified as a "bowl of spaghetti." When it came time to build a new layout, I looked for more room. The area I found was about 10 ft. by 10 ft., but in this area there was a twelve cubic foot, 36 inch high, chest type freezer with a door on top. I considered building a section that would lift up when someone wanted to get into the freezer. What about rolling stock? Visualize your wife opening such a section and having a string of cars fall 290 scale feet to the floor below! The lift-up section idea got ruled out fast. Then someone suggested building the layout high enough that the freezer door could be opened. This sounded like a good idea, and was cheaper than buying a new upright freezer for my wife! A little research showed that 52 inches would be enough. That sounded pretty high to me, but I had no choice, so away I went with "L" girder construction at 52 inches. When the frame work finally was up and a little track was laid, the kidding started. There were comments like "Do you furnish oxygen masks?" and "Do you get nose bleeds from working so high?"

After the layout had been up for a while the kidding stopped. One of the most ardent critics recently moved and rebuilt the framework of his layout. His old layout was 43 inches; the new one is now 46 inches (this was quite a change, considering that he is four inches shorter than I am). Most of the other hecklers have also had second thoughts about how high a layout should be. I even thought for a while that I might not end up liking a layout at this height. Now, after almost 2 years, I never find myself wishing that the layout was lower. Sometimes I have even thought of raising it to 56 inches, especially when I bang my head ducking under it.

Now that a lot of the track is in, much of the scenery up and a few buildings scattered around, things are starting to look like a railroad. When a steam engine goes by, you see more than just the stack, steam dome and bell; you can see wheels, valve gear and connecting rods in motion. When you sit on one of the barstools at the control panel, you must look up at a train as it passes over the trestle. It certainly is different, but nice, to see all of that underbody detail on the cars as they go rolling by. The top of the trestle is 62 inches from the floor. Even though the trestle is on a 26 inch radius, the 80 foot passenger car overhang doesn't look bad. As a matter of fact, you hardly notice it standing or sitting because you are not looking down on the car tops as you normally do, and you don't see the rail bowing out from under the side of the car. This is a real advantage to a person that likes 80 foot passenger cars, but can't afford the room for 32 inch and larger radius curves.

I never thought much about the "bowl of spaghetti" problem until one day, when I tried to take a picture of my layout while

It's all in your point of view notes the author and he provides the photos to prove it.

Photo at left shows realism gained by elevating pike to near eye level; mountain towers above scenery, yet trains can be viewed up close. Three photos at right show "spaghetti bowl" effect as camera is raised above layout.

Photos by the Author

standing on my wife's washing machine. Looking down from that vantage point was quite a surprise! The layout looked like just a bunch of track going every which way. The series of four photos showing the effect of different viewing angles were all taken from one spot. The only change between photos was the camera height.

The photo on page 24 was taken at 54-1/2 inches above the floor, which is about eye level when sitting on one of the low stools at the control panel. The top photo on page 25 was taken at 60 inches; a short person standing, or a six footer sitting on a high stool (30 inches) would get this view. The middle photo shows the view from 68 inches and would be about normal eye level for most men that visit my layout. The bottom photo indicates the view from 80 inches and would be the "normal" view of my layout if it were built 40 inches off the floor.

Still not convinced? There are additional arguments in favor of high layout levels:

Storage: The higher the layout. the greater the storage space below. Besides, it's harder to try storing things above a layout.

Wiring: It's a lot easier to sit on a box and do your wiring than to sit on a cold floor.

Kids: My small children don't bother the layout and their friends haven't bothered it either. Most of them are so short they can only see the edge of the layout. I guess the old saying, "Out of sight, out of mind," works.

Access Holes: You have two things going for you here. You don't have to crawl back to the access holes, just stoop over. Since you are looking ahead and not down, you look across the access hole. This way the access area is not seen as a hole, but as a valley or depression of land. Consequently, you don't need an access hatch.

Working: If you've ever worked on a low layout for any length of time, you probably ended up with a back-ache as a result of leaning over. With a higher layout you do less leaning. You can always stand on a box if you would like to be a little higher, but it is darn hard to dig a hole in a concrete floor when you would like to be a little lower.

Control Panels: With a higher layout the control panel can be made in more of a vertical plane instead of flat. This can actually save you some floor space.

If, after all this, you still are not sure you would like a layout at 52 inches, remember that it's much easier to cut a few inches off the legs of a layout than trying to add a few inches later!

One word about planning a higher layout. Most of us develop a track plan and think only of the top view. With a higher layout, you see more of the sides of things like hills and bridges, so keep the side view in mind as you do your planning. Everything considered, you should be able to attain greater heights in model railroading!

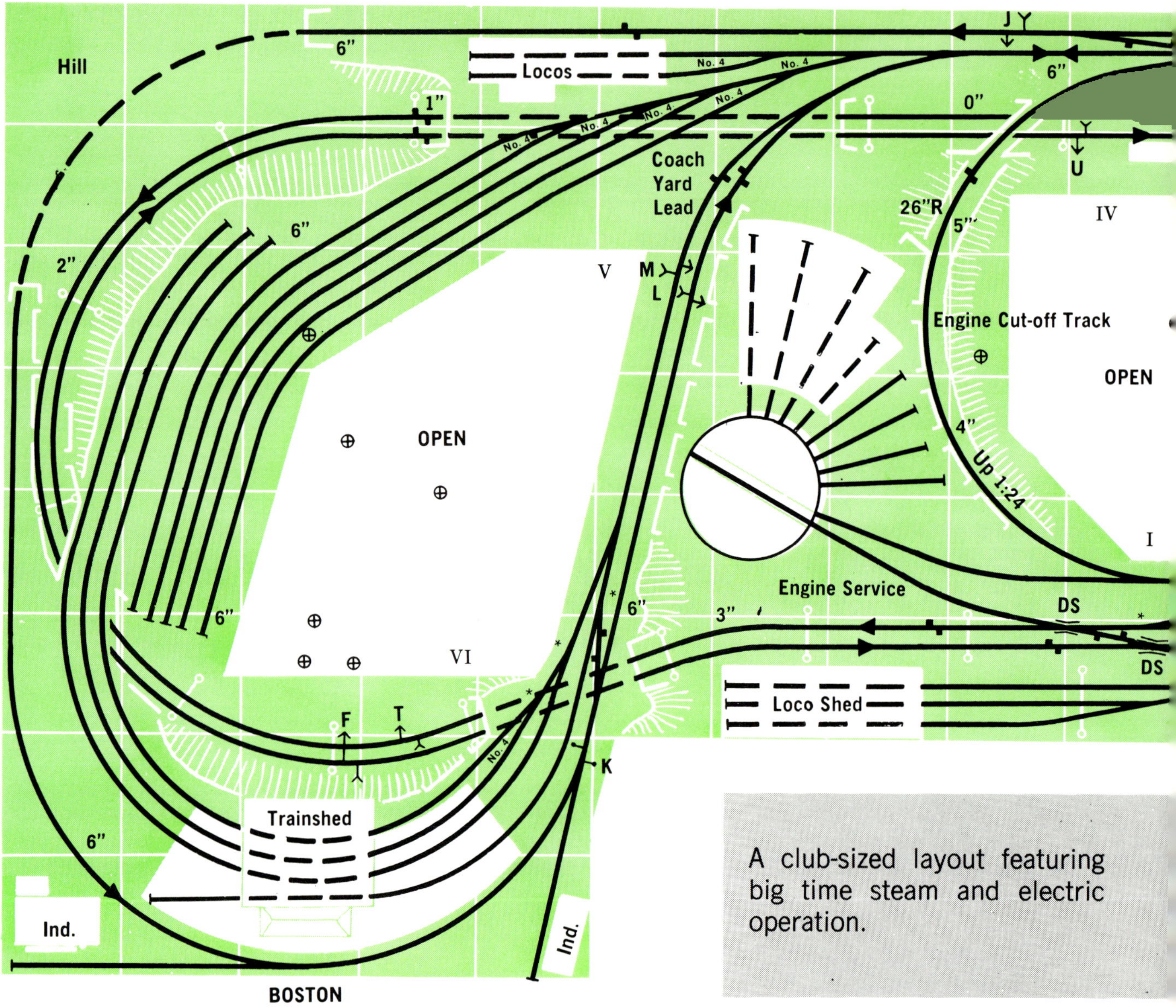

A club-sized layout featuring big time steam and electric operation.

Main Line Corridor Railroading

by the layout doctor

When my Great-Uncle Joe Detterer came to live with us during the Depression he brought along his Edison phonograph with boxes of old cylindrical records. Amongst the records, my favorite was "On the New York, New Haven & Hartford," whence cometh the tag-line to each verse "In Boston on time!" (something that may have been true in Edison's day, but not now). So when NYNH&H seemed an apt name for this electric/steam layout, that song began running through my head—you know how it is sometimes—just the tune and those last lines.

I think this 20'x8' HO layout would be a good one for a club since there can be so many operating positions besides a few engineers to slow, stop, and reverse the trains. I have circled Roman numerals to indicate possible control positions: (I) the electric/steam engine change point at New Haven; (II) the right throat of New Haven terminal and the MU storage tracks; (III) New York loops and throat; (IV) Providence crossovers; (V) upper freight yard lead and engine shed; (VI) Boston passenger station and coach storage tracks. In addition there should be some positions for engineers, power dispatchers, or whatever you want to call them. I would build a one-place control panel in the vicinity of (II) with duplicate controls so that it could be operated by one man.

Action starts at New York where there are reverse loops for MUs, other passenger trains, and freight trains. You have to use the outward main line track as a yard lead, true, which is bad. If that kept you awake, you could make the stretch through Stamford three track (slipping into two tracks to the left) and make the center one a switching lead. Operation here is by electric engines—note the catenary bridges sketched across the tracks—which could be any of the numerous electric locos now available. It would even be in order to use steeple cab locos as switchers, equipped with pantographs to contact the catenary.

Leftward from New York, on the double track, a train goes through Stamford, and almost to New Haven. Just shy of the New Haven slipswitches, the train stops, and the electric loco is uncoupled, pulling up directly in front of New Haven station, then backing into the loco shed. Meanwhile a steam engine which has been waiting up ahead somewhere (or on the turntable) comes over and backs onto the train. There is then a station stop at New Haven and the train continues to Providence. From there it

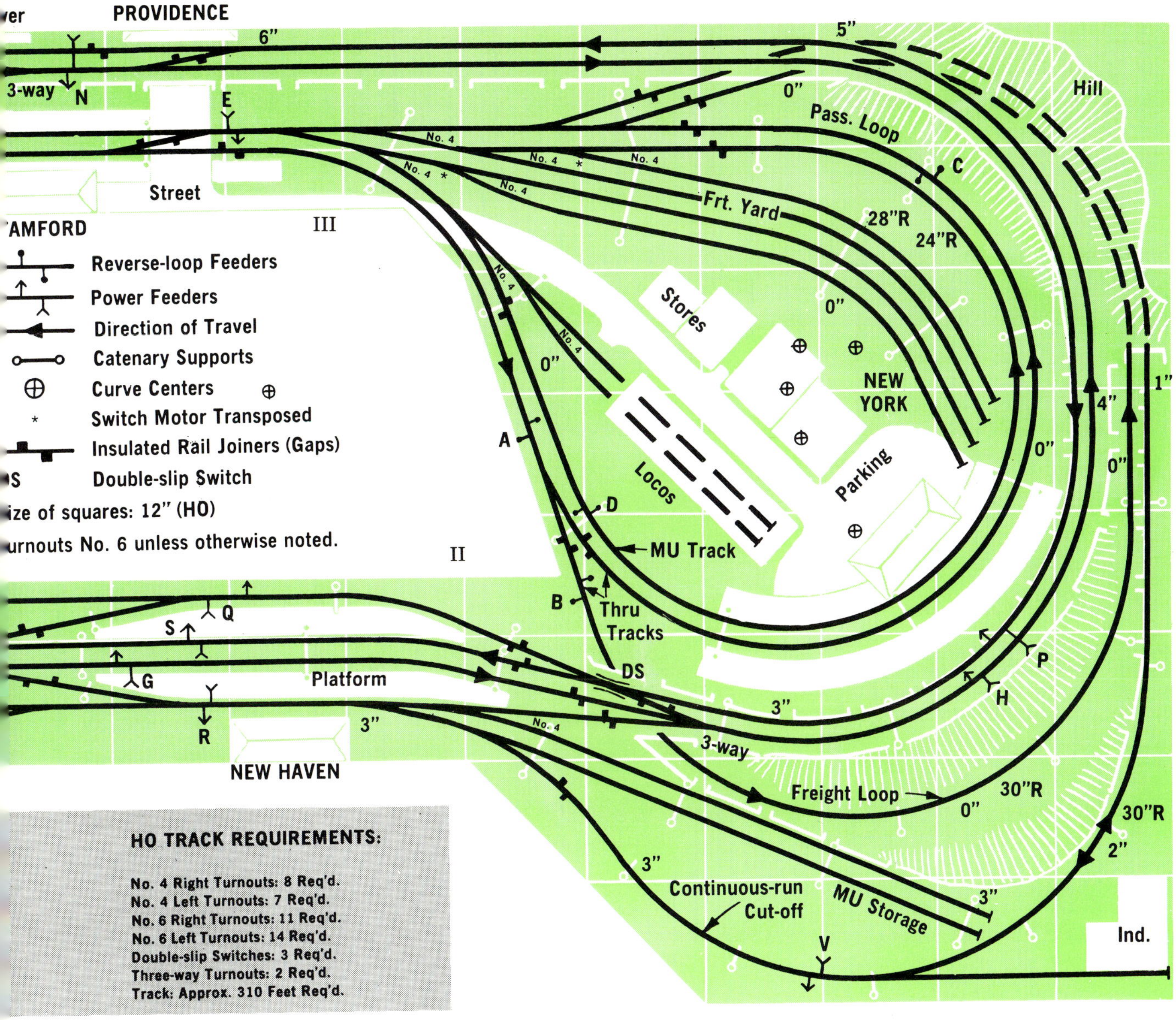

NEW YORK, NEW HAVEN & HARTFORD RR

rounds a single track loop and stops in front of Boston station. If passenger, it pulls up on block M, and backs into one of the three coach storage tracks awaiting the return run. If a freight, it pulls up on block N in front of Providence station and backs into the freight yard. Again the outward main line is used as a yard lead—bad, I must admit—but again there is space to three-track it rightward from Providence and use the center track as a switching lead.

The same system could be used in any division point set-up. Steam and diesel are exchanged at New Haven, or engines of one road for those of a connecting road. That ladder of two double-slips across a two-track main makes an ideal engine exchange set-up, with one engine house on one side of the track, the other on the opposite side. On the return trip from Boston, the train stops in front of New Haven platform while engines are changed in fashion similar to that just described. MU trains from New York all end at New Haven and are stored in the two-track MU yard to its right.

There are two cut-offs on the layout. The Continuous Run Cut-Off starts to the right of Stamford and comes around to the right of New Haven. Though it is single track, it can be used in both directions. In fact, before the track is extended to Providence and Boston, trains can make continuous runs around the "circle" between Stamford and New Haven.

The shorter cut-off, on a very steep (1: 24 or 4%) grade runs from New Haven to Providence to the right of the roundhouse. When used down-only it can provide another "circle" for trains or light engines between those towns. Its original purpose was to house Boston-end engines at New Haven, but later I was able to squeeze in a two-track enginehouse above the freight yard. Turning these engines would be accomplished by the Boston reverse loop, section K. At the New York end, engine turning can be done by any of the three reverse loops, D, A-C, or B.

A table of HO track requirements is appended. The same would be used for N gauge with appropriate changes to suit differing trackage, if built the same size. However, if built half 8x20 for N, the footage of flexible track would be halved also. If built for O scale in 16x40 ft., footage would be doubled, and an access opening would have to be provided where I've marked "stores" at the New York end.

As to the switches, I suggest that the #4s be Atlas Custom-Lines, actually around #4-1/2, with larger radius than other make #4s. The slipswitches and 3-way switches have to be Casadio, and you could make the #6 switches either make. However, where there is a star, indicating that switch machine has to be transposed to the curved side of the switch, you'll have to use Custom-Line since they are more easily transposable.

by Al Westerfield

Here's a trackplan which features plenty of scenicking possibilities, yet provides for extensive operation in a logging and mining railroad vein—and it's designed in three modules for relative ease in transporting the pike when it comes time to move.

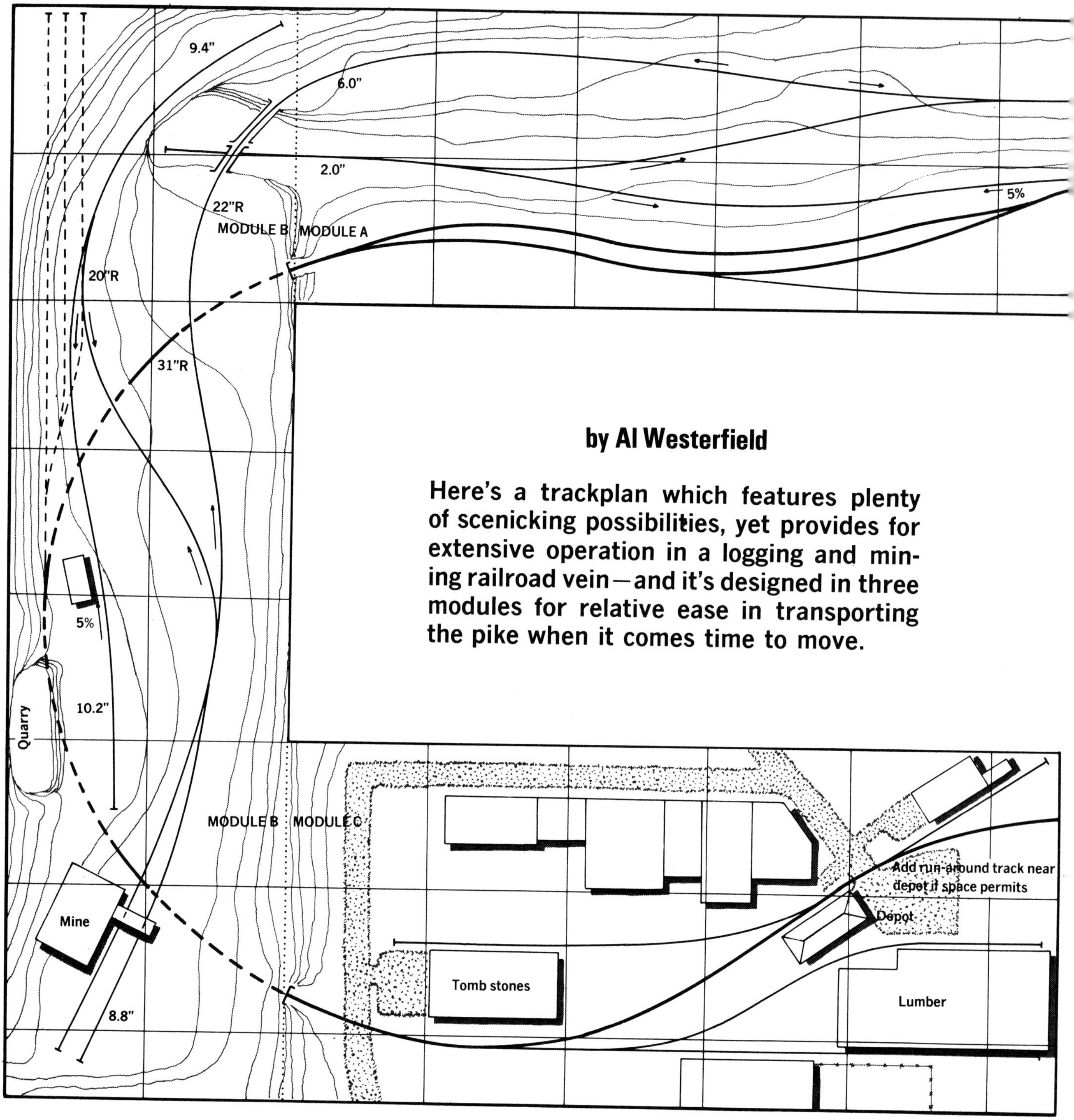

Scale: Each grid represents one square foot

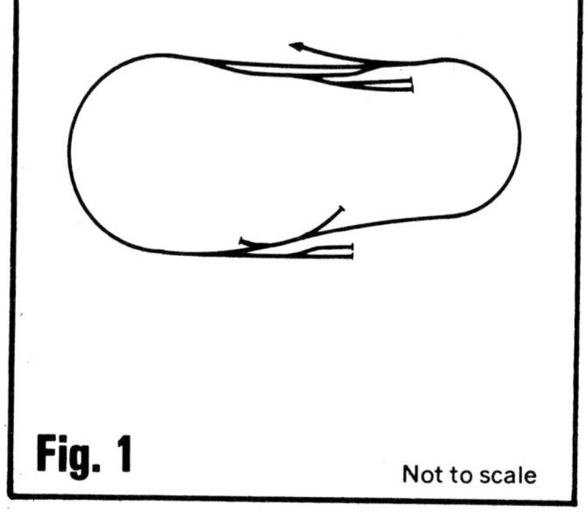

Fig. 1

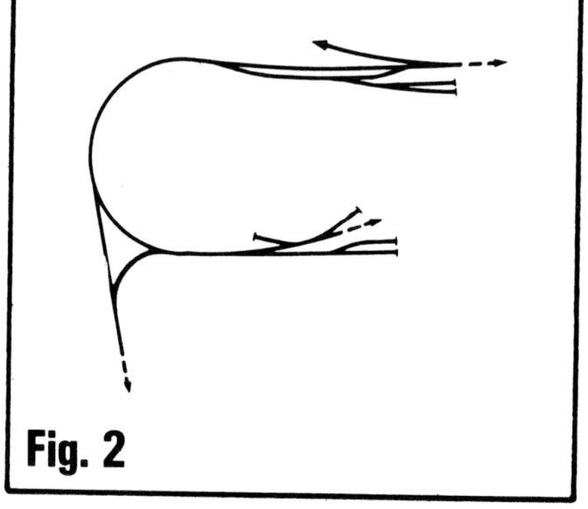

Fig. 2

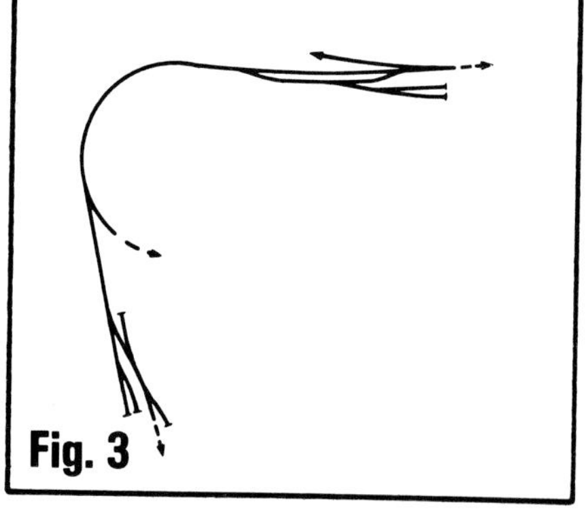

Fig. 3

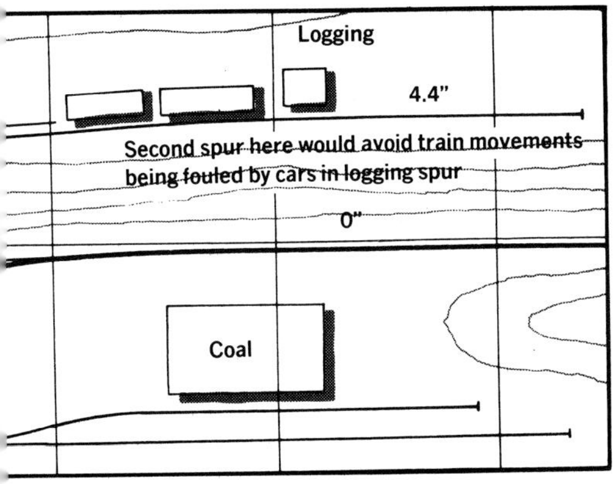

A Modular Switchback Trackplan

No question about it, the secret to a knockout layout is lots of scenery, even at the expense of trackage and train movement. And, just as surely, the secret to continued interest in a layout is trackage and train movement. All this is usually irreconcilable in a small layout. Some concessions are always necessary, and the layout shown here is no exception. If built simply "as is," and not as part of a larger pike, forget about long freights and varnish service—they would only be able to run a short distance before having to back up. If you can reconcile yourself to that, what you get is nearly a scale mile of running over extensive switchbacks in a backwoods scene.

The layout is designed primarily for HO scale and the figures on the plan reflect this fact. The switchbacks are constructed to accommodate a small engine (less than 10") and three standard length, or four to five shorter, cars. Lengthening the layout would result in accommodating more, of course. Grades are 5%, not unrealistic for either the scene depicted or the load being hauled, assuming a geared steam locomotive (Shay, Heisler, Climax) is utilized. This grade allows a 10.2" maximum elevation from the main. No. 4 switches are used in most areas. Some radii are shown on the plan; however, these are only approximations for the switchbacks which are really compound curves with easements. The minimum radius is 18", found only on the industrial spurs.

To get a better idea of the layout, let's assume we're riding a load of granite down from the top of the mountain. The quarry is a small operation with only one loading track. They can load three 35' flats at a time. Unfortunately, only two can be hauled down the first switchback at once. The mine down below is double tracked, however. Since it is railroad policy to use one of the sidings for delivery of empties at all times, two of the flats can be spotted there while the remaining one is picked up.

With the load together again, we can coast down the side of the valley, passing over a small ravine, and down to the logging camp. On the way up, our engineer noted that only two log buggies were spotted there. As a result, we can squeeze into the switchback without sectioning the train again. Had there been more, uncoupling would take place in the hollow formed by the two grades. Moving into the final switchback, we note how the railroad has utilized the terrain to lengthen the switchback by building into the ravine.

The final move is onto the main, where flat right-of-way begins. Delivery is to the tombstone works in town on the other side of the mountain. The loco makes a run-around on the passing track to be in proper position to work the industrial siding. In town the train must be sectioned again because of obstructions on the main (end of table). Similar sequences are followed for runs to the mine and logging camp. As in the prototype, logistics become important to prevent locking oneself in.

The long siding passing the coal company's loading facility is used as a drop for cars going to or coming from other destinations, necessary if the module is connected to more trackage. In this way, whole trains can be made up from the output of the mountain's industries.

Since the town must ship finished products to other destinations, hidden storage tracks under the mountain provide a source of boxcars, etc. This feature is especially important for the lumber mill which receives the logs on one track and dispenses cut boards to the other.

This switchback layout has many possibilities. Its basic size conforms to the small space of an apartment or guest room. The C-shape permits entry, with tracks running around the walls. Its modular construction permits disassembly into three pieces 2'-0" to 2'-6" wide, and no longer than 8'-6". These sections can easily be added to a larger layout in several ways. An oval is the simplest possibility; see Fig. 1. For the large layout, an industrial island might be an effective way to weave around cellar posts. A hidden wye is one way to accomplish this; Fig. 2. With a slight adjustment, the layout could be made L-shaped. As a matter of fact, if an additional track is laid carefully enough, section C could be made interchangeable—either C- or L-shaped; see Fig. 3. The L-shape can, of course, be the corner of a larger layout.

The plan has been drawn to be self-sufficient. Therefore, the passing track and hidden storage have been provided. The latter may prove unnecessary if other modules are added. Modules A and B can be constructed without module C if scenic running is all that is desired. However, I don't like the logic of running nowhere. The industrial areas provide definite destinations for each raw material.

Module B can be built with a slight scenery overlap to hide most of the joining seams. It is far easier to hide an angle gap than a flat one; see Fig. 4. The higher elevations of A and B can be mated using contoured fascia boards. One way to make a smooth transition is to cover the gap continuously by the paper towel-plaster method and then saw it apart as in Fig. 5.

These modules provide a wealth of surface area for scenicking. Only the necessary details have been indicated in the plan, as each person's tastes vary. Obviously, quite a number of small bridges could be accommodated if desired. But, beyond scenery, switching moves and the time necessary for their completion have been maximized without excess confusion. This "compromise" layout thus doesn't compromise much.

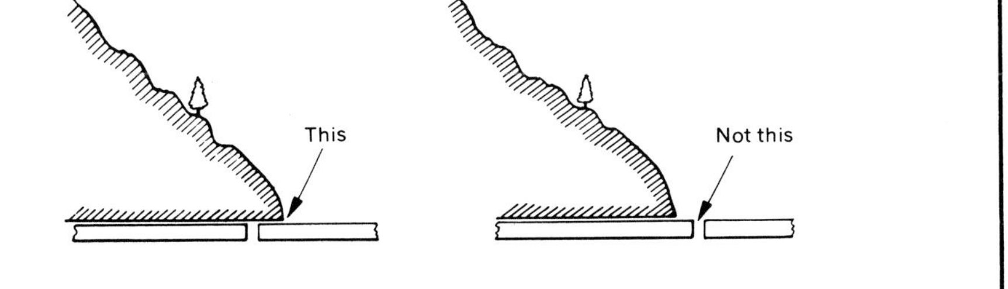

Fig. 4

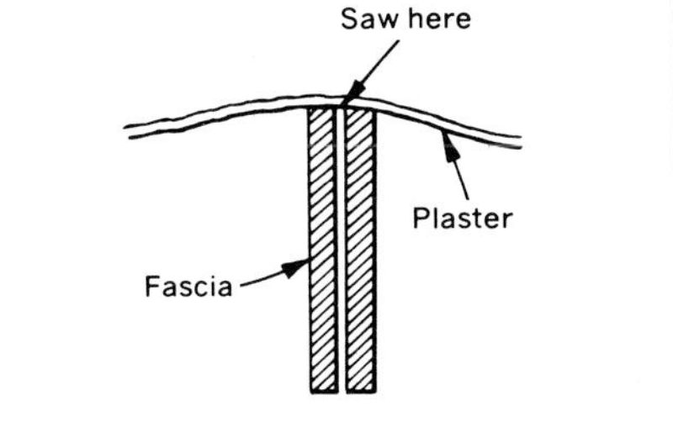

Fig. 5

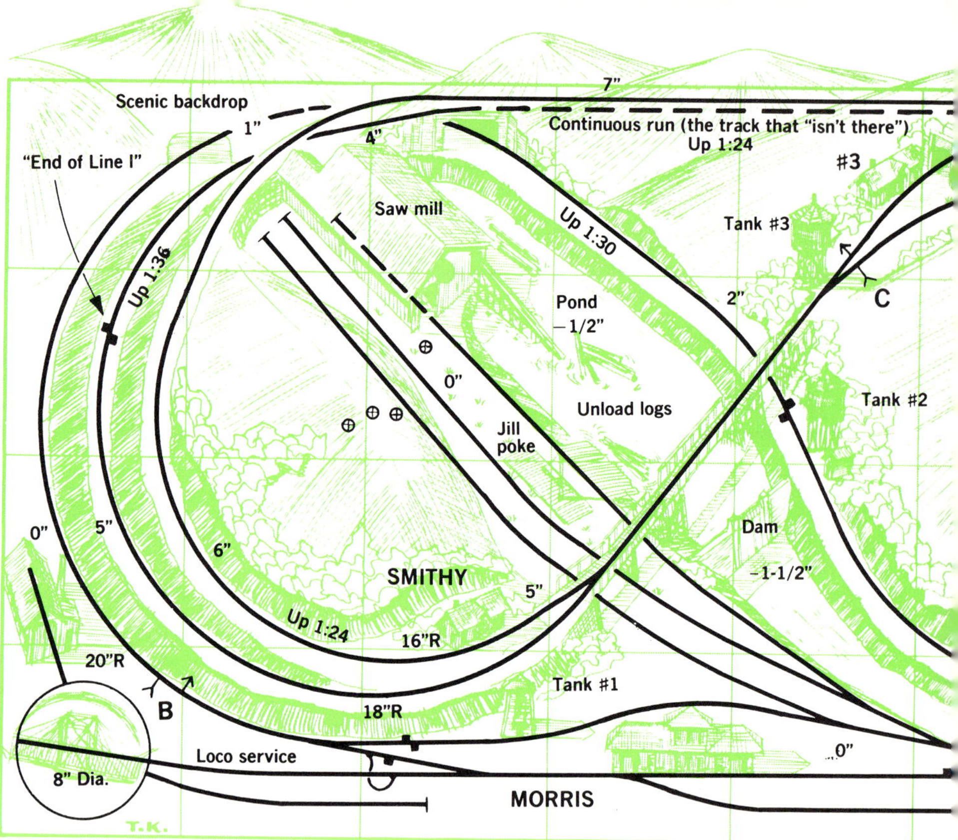

Logging Offers Curves, Grades, Trees and Trestle

TWIN PEAKS LOGGING COMPANY

With so much logging equipment on the market in the smaller scales, it appears timely for me to present another logging layout. It can be built in HO standard gauge, preferably with code 70 rail for better appearance; it can be built in HOn3 for which flexible track, built up switches, and even automatic couplers are now available; or it can be built in HOn2-1/2 for which much HO scale diminutive rolling stock and trackage can be purchased at hobby shops. I have drawn it for HOn3; at least one importer has the #5 switches which are specified. However, even #6 switches would do since narrow gauge switches for a given frog number are not as long as similar ones for HO. For HOn2-1/2, the switches might be even sharper than drawn. Indeed, if sectional rather than flexible track is used, you could contract the layout to about two-thirds of the indicated 4'x8' size.

The place to begin building the Twin Peaks is at far right, "end of line II," working around toward the front to the vicinity of feeder B. This could all be at 0" level for now. I'd suggest putting in the three yard switches, but extending any permanent track back from them until the long trestle above them is ready to be placed. From the engine house vicinity, you could continue up the 1:36 and 1:30 grade to and beyond Tank #2. Build as far as Tank #3, then hold it. Go back where you started and from "end of line II" revise elevations to make the 1:24 (about 4%) grade for the continuous run "track that isn't there." Bring this around front to Smithy and then put in the trestle. Before you put it in, you can complete the yard tracks. Last would be the branch which runs down the back, behind the scenic background, and at right center, above two other tracks.

The twin Peaks Logging Co. can be operated as a point-to-point line, from "end of line II" to "end of line I." The "track that isn't there" would be considered in this case as truly not there. "End of line II" at right would be considered either the outer end of the yard lead, or a connection with another railroad, perhaps of a different gauge. "End of line I" at left would be the upper end of the main line, but the branch would extend farther on. When operation of this point-to-point railroad palls, or a continuous run is desired for exhibition, relaxation, or running-in, the "track that isn't there" could be put to use. Since it is on a 4% grade, it might be used "down" only, or engines only, or short trains could run up the hill.

A log train entering Morris terminal at the bottom of the hill would pull behind the station. If the road engine were not to do the switching, it would run in front of the station to the turntable. Then a switcher (lurking on the coach storage track or in the yard) could come out and shove the loaded log cars alongside the pond where the jill-poke would knock the logs into the water. The cars would then be put into one of the storage tracks.

Later, empty log racks would be brought out and placed behind the station, and a road engine would pull them up the grade to the trestle; that would be as far as the road engine would go. It would uncouple on the 18"R in front of Smithy, and pull back to "end of line I," while a branch line loco would come from the loco spur and

shove the cars up the branch. Since this runs behind the scenic backdrop, you could have the layout table out from the wall a few feet so that an assistant operator could take the logs pushed into the pond (dry!) and hand-load them on the log cars again.

Before the branch loco brought down the loaded train from the branch, the road engine would have run light across the trestle, waiting past the station at #3. The branch engine would leave the cars on the trestle and pull into its spur. The road engine would couple to the train and head off down the grade to Morris. If the caboose had to be gotten to the back end of the train, the switch at Smithy would let this be done.

Holding engines as just outlined above, on the main line and then on the spur at #3, will work out nicely with the all-rail frog HOn3 switches available. The HOn-2-1/2 switches I've examined do not have this feature; hence, all rails are live. This would require wiring in short killable sections of track, insulated on one rail from the rest of the main line, and energized or killed by means of any convenient SPST (on-off) switch.

Electrically I've divided the line into three sections: A,B, and C. This is for the convenience of controlling more than one engine at a time. If you have but one engine, you should put in the necessary insulated rail joiners, but run all three pairs of feeders to one set of controls. Eventually three full sets of controls (A,B, and C)

Layout Symbols

Size of squares: 12" for HO Scale

All switches are Number 5

Insulated Rail Joiners (Gaps)

Power Feeders

Centers of Major Curves

0" Elevation of Track

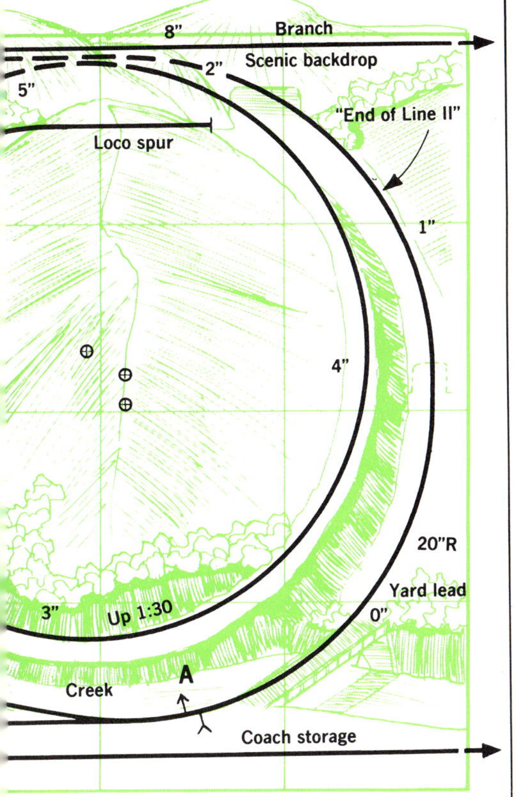

Though designed specifically for HOn3, this logging pike can be built with little modification in HO, HOn2-1/2 or any popular gauge.

should be wired to one power pack.

For motive power in HO or HOn3, I'd suggest two geared engines such as two-truck Shays—one to switch the yard, the other to handle the branch service. For main line use, you could have either a three-truck geared engine or a small rod engine such as a 2-8-2T or 0-6-0T. Something like a rail bus or Galloping Goose would be nice for passenger service, plus a few old open platform coaches for excursions.

For HOn2-1/2 you would have to use the small 0-4-0T locos now on the market, double-heading them as needed. It will probably not be long before some clever fellow has revised a Balboa class A or PFM B-2 Shay for HOn2-1/2. (Just cut a piece out of the axles and splice them with tubing, making and installing new bolsters; it doesn't look too good from the bottom, but it will work. The mechanisms of some N gauge six-coupled locos can be reworked with new superstructures.)

Track parts required are as follows: For control block A, four #5R switches, and 23' of flexible track. For section B, two #5R, one #5L switches, 18' of track. For control section C, including the branch, two #5R switches and 27' of flexible track. Since all the switches except one (to the loco spur) are within easy reach of the front of the pike you might want to use hand operated switches for the most part. Even the loco spur switch could be hand-flipped if you had a second operator behind the scenic backdrop to load logs on the cars on the branch.

An Unplanned Layout

The simple oval packed with most train sets can quickly turn into a real railroad system having a mountain division, yards, and a mining branch.

For the fellow starting out in model railroading with a train set or equivalent, the usual first layout is an oval, where engines and cars can be let out and given good long runs. This is naturally so, since an oval of track comes with every train set, albeit with mighty little straight for the sides. In all instances the curves are of the standard radius for the gauge, in HO 18" radius, making a circle 36" in diameter to the center lines of the track, somewhat larger over the outer ends of the ties. In addition, for what it is worth, I recommend the oval for a first layout to get the feel of model railroading as to speed and direction control, uncoupling, coupling, derailments and various other challenges.

The dealer from whom you buy your set will be darned happy to sell you extra pieces of straight to make the oval longer. He also offers larger 22" R curves to make a more generous sized first layout. 22" R curved track comes 16 pcs. to the circle, whereas 18" has 12 pcs. to the circle. Maybe you'd like to try using flexible track, which usually comes in yard lengths instead of the more easily portable sectional track which comes with a train set. Sectional and flexible track can also be combined in the same layout thus avoiding waste.

In the "unplanned" layout shown here, I picture a fellow who has a 4' x 8' table on which he wants to erect a maximum radius oval, using 22" R curves which will allow nice clearance on each side. It could be made in HO with 6 pcs. flexble track, or with 16 pcs. 22" R sectional track and 10 full and 2-3" pcs. of straight (of which one would doubtless be a terminal section, one a rerailer). Divisions are not shown for sectional track.

A layout this size will provide trackage to test out all your rolling stock. 22" radius in HO will handle all but a very few steam locos available in the gauge (including most articulateds), and all diesels and electrics too. If you want, you can include a switch such as the one No. 6L switch at front, and let a siding trail into the center of the table more or less as the 2-track yard is shown. (The switch should have an insulated frog, such as Atlas Custom-Line, Snap-Track or Casadio) Such a spur would let you store cars not in use and give you practice in switching and coupling and uncoupling.

Now of course any layout for which a plan is given is not really unplanned, hence the quotation marks around that word in the title of this article. However, my intention is to show you a few of the ways a simple oval layout can be expanded without too much difficulty. You'll note that if 24" of track at the front of the first oval (in color) are removed, two switches can be substituted—a No. 6 left, and a double slipswitch. With these in place (perhaps you have planned enough to have a 24" long piece of track along here to begin with) tracks can be run both sides of the main line.

You could run another oval of 20" radius around inside the first 22" radius one and have a double-tracked oval, and have two trains run at one time—under separate control if you have placed twin insulated rail joiners (indicated by black blocks) in both rails between switches of the crossovers at the front.

But let us consider a second level of track so that some interesting scenery can later be built. Let the 20" R to the left start to rise. It goes up a grade of 1" in 36" (less than 3%) to achieve the needed 3" elevation to cross the original track at right rear. That might be enough for a while. However, note how the original 4' x 8' table (shown dotted in color) can be expanded by 8" at the right and a foot at the lower right. Then the rising track can cross the main line and drop down again as drawn to rejoin the front end of the double slipswitch. There is not as much track length for this hill, so it has to be 1" in 30" which is more than 3%. So this new line rises to cross the original line and drops down again. You don't want to talk about this too much because it seems pretty silly, but really no more foolish than having a train chase 'round a continuous track, nor many other things that iconoclasts proclaim about model railroading.

This can be called a "mountain division" for you can now start to picture it winding around among the hills. But yet another addition can be made to the trackage in this "unplanned" pike. How about a small terminal? Add 3½ ft. to the lower left, about 1 to 2 ft. wide as you can spare space, and you have a terminal site. This should be 3" higher than the previous trackage since the "throat" of the terminal will have to cross above earlier track—opportunity for a scenic tunnel entrance. The tail end of track to the right, inside the main line, can swing around 20" R, later 18" R to snake up a 1" in 36" grade to a terminal. It's not so much now, but with a longer added platform it could have more tracks. If not as much width as drawn is available, the tracks could be straightened out.

In this added area I show a town for the branch to serve, and as an afterthought, one of the HOn2½ industrial narrow gauge lines climbing higher.

This "Bridgeton & Saco River" as I call it is not intended to be your final layout, but is intended to show how a pike can be expanded. A great many fellows get their model railroading kicks out of constantly changing and rebuilding their model pikes. I know of one man, now in the hobby business himself, who

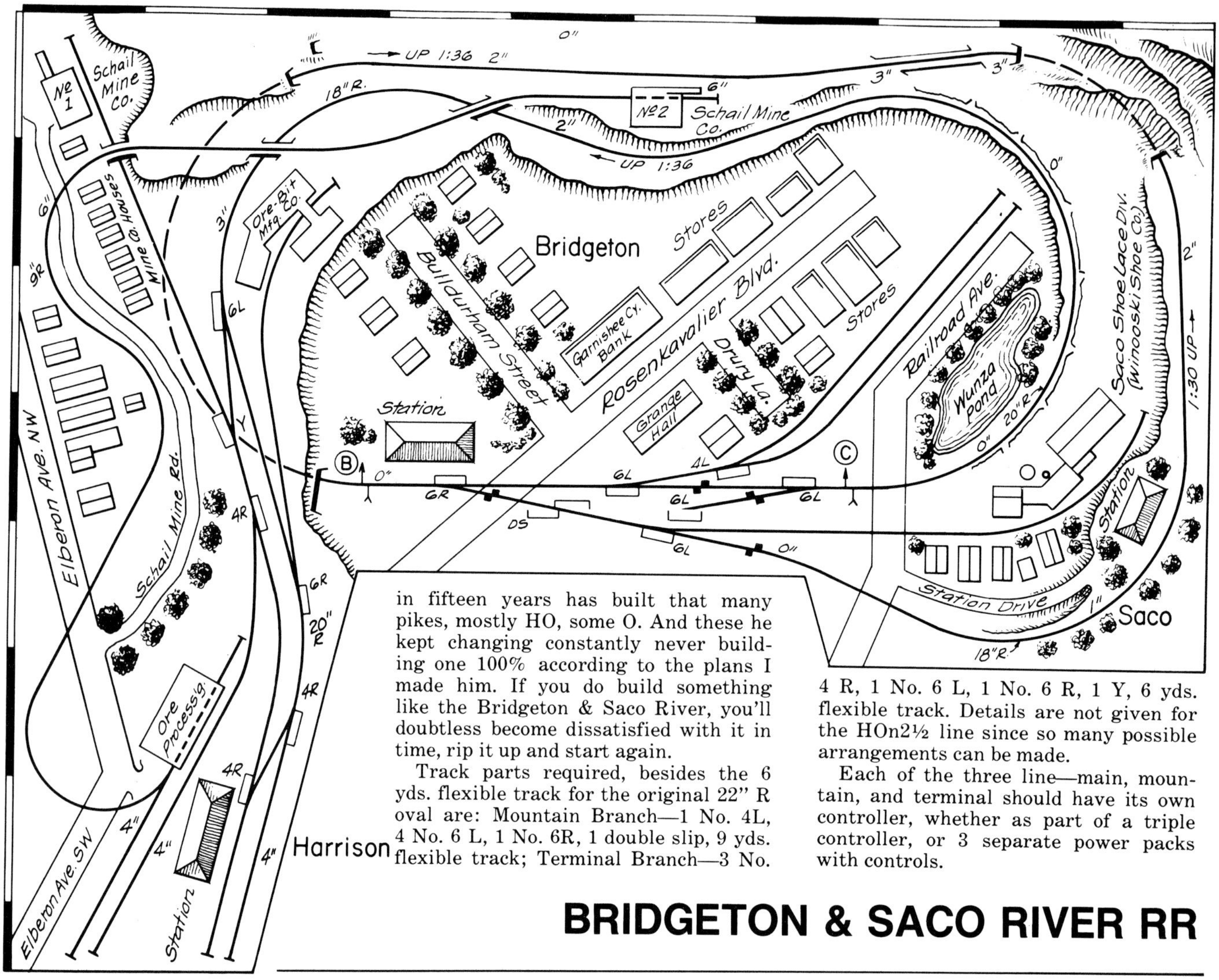

in fifteen years has built that many pikes, mostly HO, some O. And these he kept changing constantly never building one 100% according to the plans I made him. If you do build something like the Bridgeton & Saco River, you'll doubtless become dissatisfied with it in time, rip it up and start again.

Track parts required, besides the 6 yds. flexible track for the original 22" R oval are: Mountain Branch—1 No. 4L, 4 No. 6 L, 1 No. 6R, 1 double slip, 9 yds. flexible track; Terminal Branch—3 No. 4 R, 1 No. 6 L, 1 No. 6 R, 1 Y, 6 yds. flexible track. Details are not given for the HOn2½ line since so many possible arrangements can be made.

Each of the three line—main, mountain, and terminal should have its own controller, whether as part of a triple controller, or 3 separate power packs with controls.

drawn by James Berge

BRIDGETON & SACO RIVER RR

CALIFORNIA & SOUTHERN RR - HO - 9-1/2′ x 22′

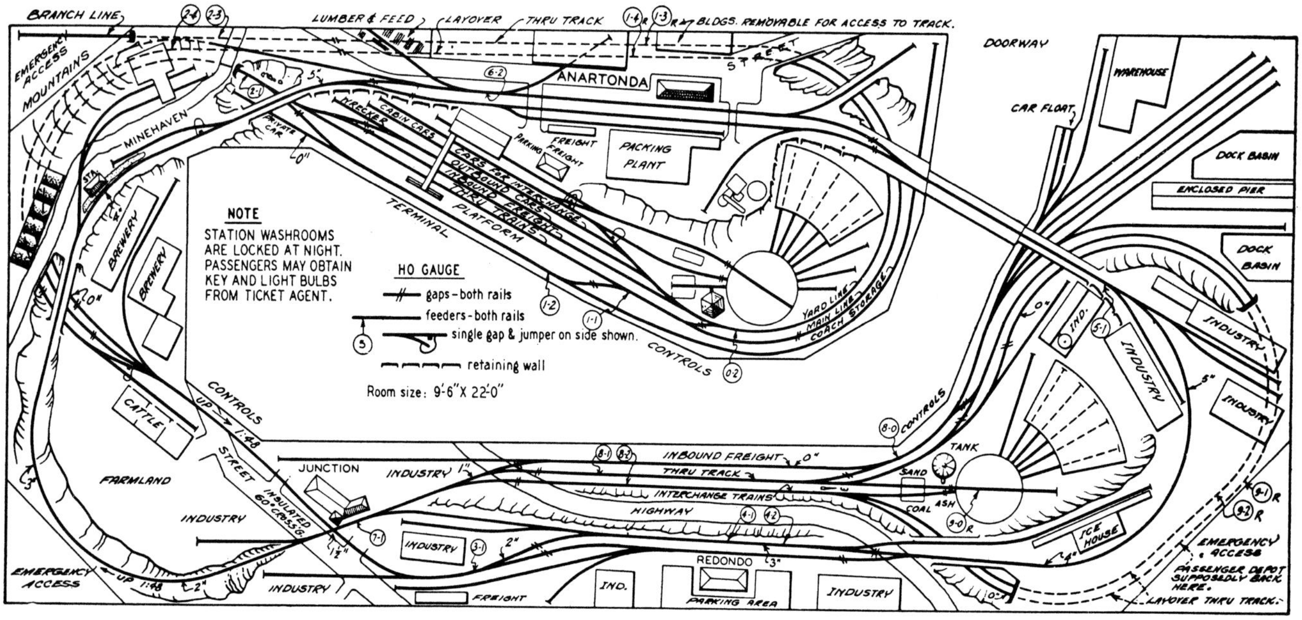

Here's a large HO gauge layout that's received an official okay from the Layout Doctor. Designed to completely occupy a large room, 9½ x 22 feet, this layout will provide diversified operation and will keep three good men busy. Note that most switches are relatively easy to reach although some are located three feet from an open area. On first glance, this layout appears to be complex, however if built in sections, it would be relatively easy although a time-consuming job. The finished product would be well worth any effort involved. Construction should start with a complete loop around the room, in an effort to get equipment into operation in short order. After this loop is installed, the various switches, sidings, buildings and scenic details can be added as time and money permit. Track elevations are shown and these should be followed carefully to get the proper clearances at tunnels and the correct percentage of grades. Note that this layout includes residential and industrial areas, mountainous country, farmland and even port facilities. There are many sidings, industrial spurs and large yard facilities to provide interesting operation. If space is at a premium, this pike could be shortened in length by at least six feet by sacrificing radii and grades slightly.

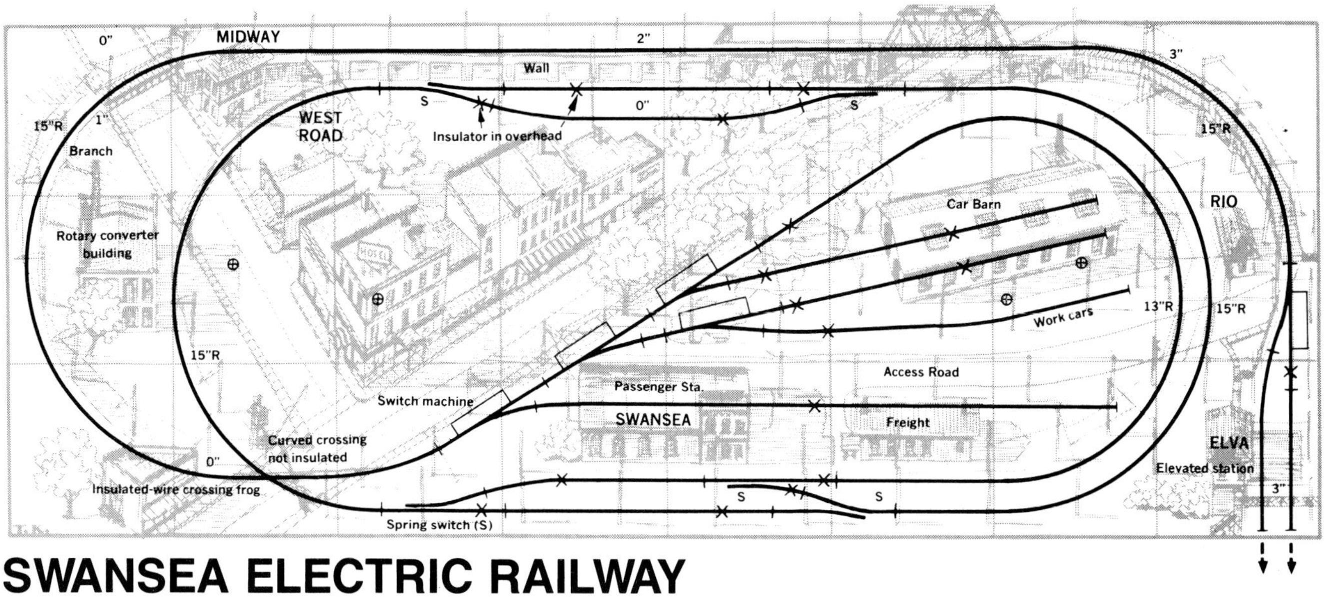

SWANSEA ELECTRIC RAILWAY

Beginner's Traction Pike

The Swansea Electric Railway (3'x8') has several features designed to let the new traction buff get his feet wet in both track and overhead wire construction.

by the layout doctor

The fellows who crave a huge cellar-filling model railroad with multiple lines and spacious terminals and yards, yearn for the ability to make up and dispatch several trains at a time—for real railroad action comparable with, say, one of the Penn Central main lines. Yet if a steam/diesel railroad modeler can generate an interest in traction, he can get as much action in a fraction of the space: one-car "trains" combining engine and cars in one "unit"; sharper curves; sharper switches; steeper grades; and the general reduction in layout size that these factors permit.

What is traction? It means over-head wire or third-rail operation, whether city street-cars, interurbans, or sometimes heavy electric lines such as parts of the Milwaukee, PRR and NH. Some widely sold brands of HO traction equipment will run two-rail (some HO, as well as O equipment, is intended from the start for overhead running) and actually must be rewired to operate from overhead wire—not too hard a job.

As one example of what can be done in HO traction, I show the 3'x8' Swansea Electric Railway, a slightly longer version of my own first strictly trolley-operated line, with improvements. It is just a single tracked oval, with a passing track at front, another at the back, plus a small terminal, barn and branch line (the latter shown rising to terminate at a small elevated station). The single track oval part has 15" radius curves. It would be best to build the switches up, using all-rail frogs, but I have drawn it to utilize regular Snap-Switches, which have insulated frogs, so that you would have to *un*-insulate some wheels to avoid slow speed stalls at the frogs. The Snap-Switches can be set for one direction, yet heavy cars can easily trail through the sprung points from the other direction. Five of the switches show no motor housings, though you will need to spike down some steel piano wire lengths to spring the switches. They are always set for directions indicated on the drawing, but cars can trail through them from the direction against which they look to be set (these are *spring switches*, quite common on real electric lines, and extremely useful on any model traction line from the standpoint of economy, wiring, and perhaps even derailments). A model traction line can be designed as shown here with all spring switches except where there really is a choice of routes as at the barn, terminal track or station track on the elevated.

The curved-over-straight crossing (lower left in the trackplan) is the most difficult problem facing the builder of this pike. Even if you have the crossing custom built, a much better definition than "straight track and 15"R curve intersecting," is needed, since there are an infinite number of "angles" possible ("angle" is not best way to define it). Your easiest out is to *build* the crossing yourself, and *not* on your workbench but right on the spot. What makes this relatively easy is that the crossing does not need to be insulated; it is not two-rail.

The 15" radius curve and "loop" around the barn should be installed first, then brought up to the inner rail of the curve, rails fitted against each other, and soldered in gauge. With the help of a ruler, the short straight lengths between the curved running rails are shaped with ends beveled, then spiked in place. The straight on the left end is then added, filed to shape, solder-

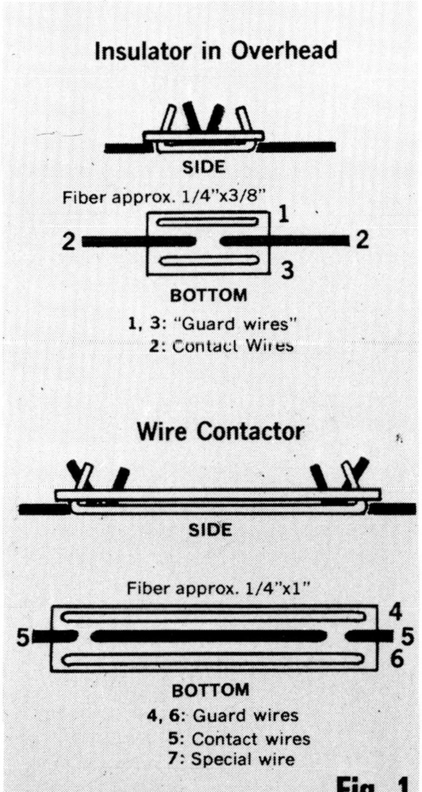

Fig. 1

ed to the outer curve rail, and guard rails positioned as needed. The four frogs can be filled with solder, filed smooth, then rutted out for flangeways with a warding file or a separating disc in a hand motor tool. Refer to the sketch of this work, with steps numbered (Fig. 2).

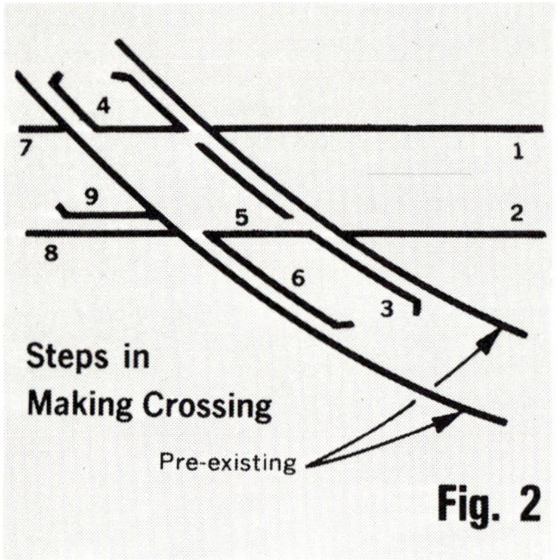

Fig. 2

When I had a similar layout, my branch line stayed level around to the station at Elva, suggested to be elevated in the manner of the Sioux City Elevated or one of the older and long-gone New York els. The arrows below suggest that this trackage can be extended down along the wall to a new area of the layout some day.

This is not a treatise on hanging trolley wire, but I will remark that it is easy. Bowser makes a nice set of bracket-arm trolley masts, with a coil of suitable bronze contact wire and special clips for doing it all without soldering. This makes up into a good looking and good running over-head wiring system; on "double track," separate poles with brackets for each track are required, unusual but not unheard of in prototype. You can also use 1/4" wooden dowel or 1/8" steel, brass or copper rod for masts. I prefer stretched #20 copper wire for overhead since I used it on all my lines, but it is bulkier (though less springy) than bronze.

For non-Bowser overhead, there are several makers of lost wax hangers and such available. What is not currently available to my knowledge are insulated wire joints and contactors. Fig. 1 shows a wire joint of thin fiber with inserted "guard wires" and the two insulated bits of contact wire all hooking in. The wire contactor, used for signaling, etc., is also shown, about the same design but with a separate length of contact wire in the center. I used to use such contactors not just for signals and control, but for throwing switches: if the trolley kept going, switch stayed as is; if it stopped, then started with pole on contactor, and the switch would throw.

By the way, much of the overhead material made and sold for O gauge traction is also used in HO, and vice-versa, so do not shun hardware because of "wrong scale".

For the Swansea Electric Railway, you'll need two left-hand Snap- or hand-made switches, eight right-hand, 45' of flexible track—and that crossing you'll make yourself.

All in all, it is small enough for a relative beginner and can be built largely with commercial track components; the need to build one piece of complicated trackwork insures that even a beginner will have the "opportunity" to learn skills needed for more advanced traction layouts.

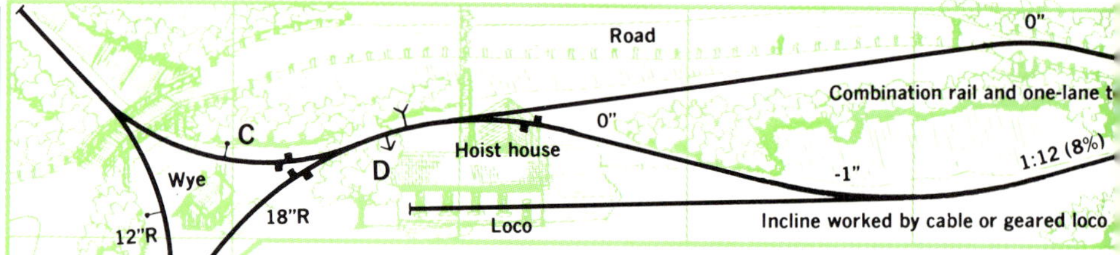

An L-Shaped Shelf

CARBON COUNTY RAILROAD

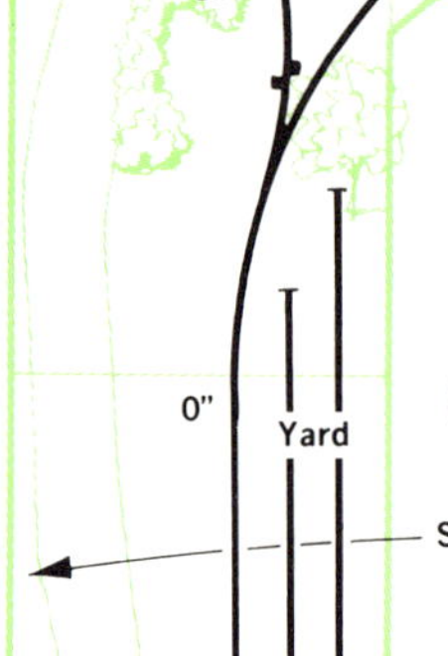

by the layout doctor

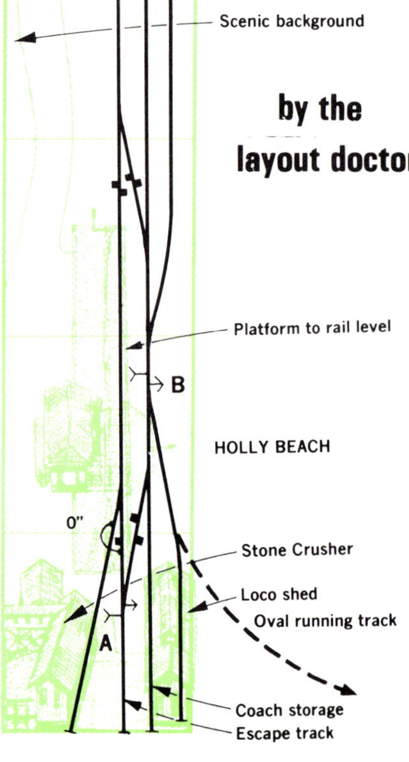

Layout Symbols

Reversing section power feeders
Power feeders
Insulated rail joiners (gaps)
0" Track elevation

Track Requirements

8 #4 right-hand switches
8 #4 left-hand switches
Approximately 36 feet of track

The L-Shaped Room, per movie title, is not too prevalent as a site for building a layout, but within any room an L-shaped space can often be utilized for a model railroad. My own previous attempts at L-shaped layouts have been L's with legs wide enough to swing a decent radius curve 180° at the end of one or both legs. However, the pike to be described here, the Carbon County R.R., is designed for an L with narrow shelves, neither wide enough to swing a radius. L-shaped pikes such as this are often featured in British model rail magazines. Aside from "proprietary" pikes—which we would call RTR—little L-shaped branch lines could almost be said to be typically British, so far as one can tell from the magazines.

Layouts of this shape, such as the Carbon County Railroad, look like a lot of fun to operate. They are really point-to-point, like a real railroad. At each end of the rather short run, the loco must be uncoupled, run around the train, turned and serviced in the case of steam power, then coupled to the other end of the train for the trip back—real railroading, by gar! However, many modelers will find the inability to run a train continuously (for testing, break-in or just to watch the trains rumble past) a drawback. Unless the builder is an experienced modeler, and therefore can be expected to have a keen appreciation and understanding of his own taste in railroading, building a point-to-point pike with no provision for continuous running may result in a drop in interest in the long run. To put it another way, most modelers start out with some sort of operation which involves a lot more train running than an attempt to duplicate prototypical operation (with various classes of trains, some sort of system for dispatching trains such as CTC or train order, perhaps timetables, etc). This too can grow very boring; one lap of the layout doesn't have much more promise of excitement than the one preceding it.

Several years in the hobby (and several layouts later,) the modeler often finds a particular type of prototype operation that he most admires, be it four track main lines, traction, period operation with small locomotives and short trains, narrow gauge or perhaps short or branch line operation. This latter group will find the point-to-point layout almost ideally suited for their purpose, and the L-shaped design is in turn well suited for point-to-point pikes. Small layouts like this often result in time to build fine scenery, another important attribute.

The Carbon County was originally conceived to be a branch line in narrow gauge or trolley, connecting a by-passing main line with a resort or town that was somehow missed. Examples abounded in the U.S., and some even remain today: the Stone Harbor R.R. connecting the A.C.R.R. with the resort in South Jersey; the Susquehanna & Eagles Mere connecting that resort in Pennsylvania; and the Catskill Mountain R.R. in New York State. A study of an old railroad Atlas or *Official Guide* will disclose many more.

The original idea in making the road HOn3 or HOn2-1/2 was not to have it just a branch of some big road, but this struck

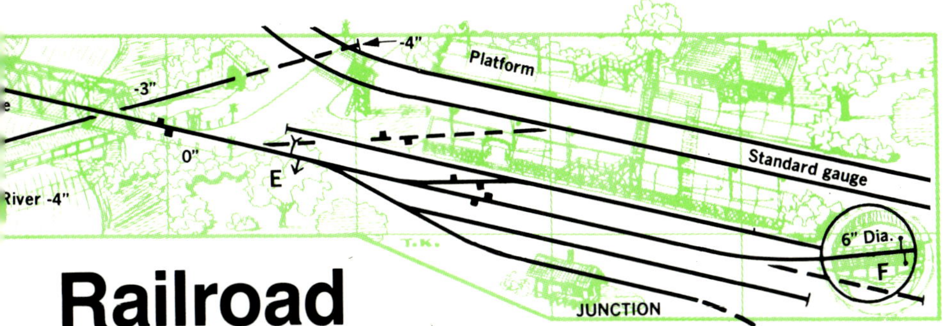

Railroad

A narrow L-shaped shelf is all that's needed for this standard or narrow gauge point-to-point layout which features lots of switching and scenic opportunities.

a snag as will appear, so it can be standard gauge, or perhaps trolley operated.

At the Junction, there is a cross-platform transfer of passengers from standard gauge to branch trains, plus a small turntable to turn the Carbon County engine. There is single track across a planked-over, privately owned bridge across an impassable ravine. This, upon payment of toll, also serves as a highway bridge, single lane only. A train rolls by Wye and soon is in the station at Holly Beach, the resort. Engines at this end are turned on the tight wye which is located in the corner.

In order to add a little operational excitement, I added on the long side a very steep (1:12 or 8%) incline to be cable operated, possibly with the assistance of a small loco. Half-way up the incline there is a covered engine shed adjacent to a power-house for the cable hoisting engine.

But what does the cable road bring up the hill? Gravel? Ore? I'll leave it a bit vague, but something has to be done with it. Hence, I added a "stone crusher" at Holly Beach. That's where the little cars from the bowels of the earth go, but then what? Is the product of the crusher then shipped out by truck? Could be. Another thought would be to add the dashed track connection, assuming the same gauge, at Junction so that loaded cars of "the product" could be shipped around the country.

With the usual oval layout, all the operating can be done from the "front" and/or from a central access space; this is not the case with an L-shaped railroad. If it is as elaborate as the Carbon County (which is fairly basic), it must operate from several points: Holly Beach; the wye and incline; and Junction. This means having several operators or having one who is mighty agile, assisted by a well-greased swivel chair. It might lead to the installation of three sets of controls or one control on a flexible lead.

I note that the British often overcome this "problem" by having one end of the L-shaped pike medium complex, with the other down to what they call a "fiddle yard" where you fiddle around. Or they might have a giant turntable on which a whole (short) train is run and turned. Except that it takes up more space away from the wall, one could have a reverse loop at the outer end, a terminal, such as Holly Beach at the business end. The British have the advantage of being able to use double-ended tank locomotives which do not need turning; this is less of a problem with diesels. Also, I believe many of their one-horse branch lines *were* operated on the push-pull principle, as many commuter trains are now operated in the U.S., except that they would have just one car with an engine. As noted, point-to-point operation would be a little easier with diesels, and even easier with gas cars or RDC's.

To build the Carbon County, 8R, 8L #4 switches and 36' of flexible track are needed, no matter what the gauge, if the 5'x10' size is retained. This does not include any track for the ghost trackage all the way around the space as indicated by the dotted 18"R curves at Holly Beach and Junction, which could provide for continuous running for those with more space.

Spiral Reverse Loops Solve Grade Problems

Spiral reverse loops provide the means of creating unusual scenic and track formation not otherwise possible in limited space. The California & Midland was designed as a 6′ x 14′ HO layout and offers room for high trestles and bridges.

The California Midland Railroad utilizes an idea which is not seen too often on model railroads, whether lay-your-own or sectional: the spiral reverse loop, shown in fig. 2. Instead of the once around, on the level, reverse loop we're accustomed to seeing (fig. 1), the spiral reverse loop wraps around an extra time, crossing itself somewhere enroute. Such a loop gives scenic possibilities to the area since where there are two levels and a bridge, there can also be mountains and valleys.

The spiral idea can be with a bow knot track formation, too, and in two different ways. In fig. 3, I show an outside spiral, where the inner loop of track is higher than the outer loop, permitting the placement of a hill or mountain in the center. In fig. 4 I show the inside spiral, with the inner loop lower than the outer one, thus permitting a lake—or a city, or a roundhouse, etc.—in the middle. Also, the inner and lower track could be in a tunnel, under a mountain, giving a greater change in track elevation than would appear possible.

To work out this spiral idea further, I've augmented the bow knot at each end into a spiral—an outside spiral at left, an inside spiral at right. It is not pure, being complicated by a cut-off track across the back, which switches in in accessible spots at the front.

Let us study this California Midland R.R., starting a train out to the left from the five track yard. A 1:48 or 2% grade starts at Terminal City and climbs past Weedland. Just across

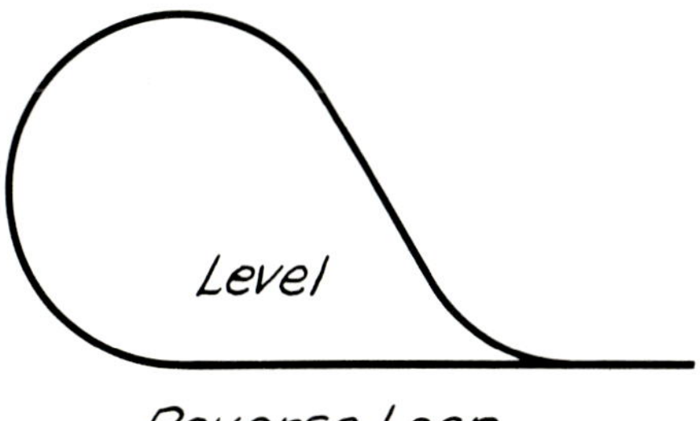

Reverse Loop

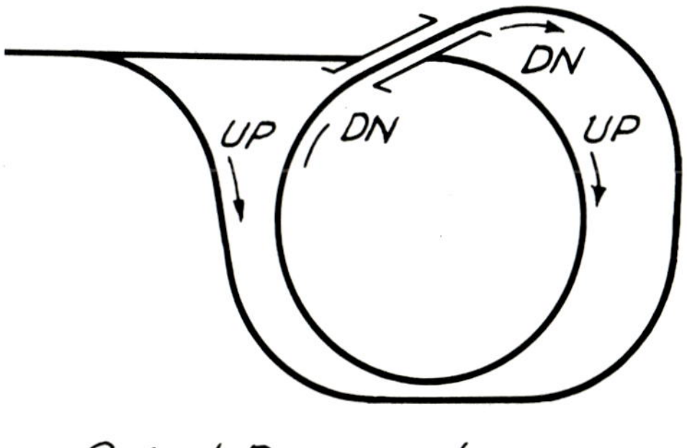

Spiral Reverse Loop

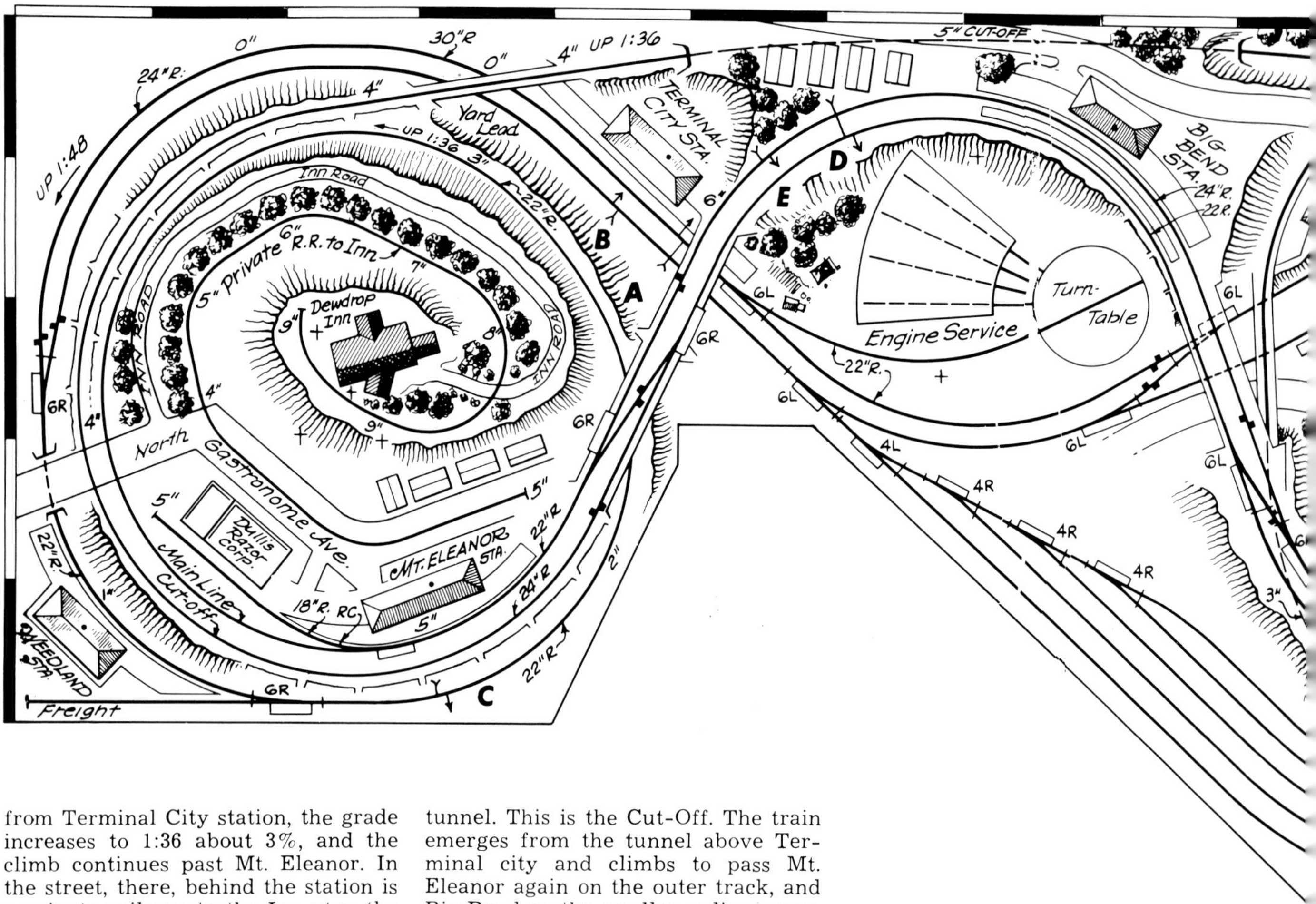

from Terminal City station, the grade increases to 1:36 about 3%, and the climb continues past Mt. Eleanor. In the street, there, behind the station is a private railway to the Inn atop the hill. This can be a simple four-wheel trolley, or it can be HOn2½ Mini-train or Jouef with an engine shoving a small passenger car up to the Inn.

Past Mt. Eleanor, the track crosses a long bridge, and on that bridge is a crossover. This didn't used to be possible when we had to put switch motors "underground"; but here you can use either Casadio or Custom-Line #6 switches nicely. In fact you can use either kind anywhere except a few spots where the larger Custom-Line motor housing will not clear another track close-by. You can also use all-rail frog switches except for the curved right switch at Mt. Eleanor station .

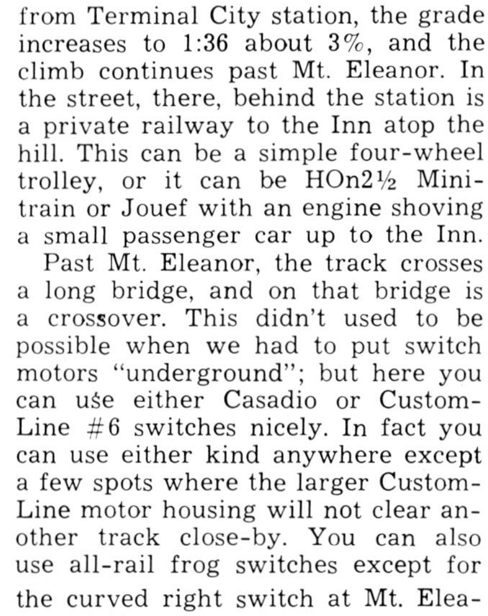

Outside Spiral

The train next swings past Big Bend station on the larger radius track, then slopes down a 1:48 grade and across a lower track into a long tunnel. This is the Cut-Off. The train emerges from the tunnel above Terminal city and climbs to pass Mt. Eleanor again on the outer track, and Big Bend on the smaller radius curve.

Thence it continues down a 1:36 grade, ducking under the Cut-Off, passing Winooski station, and into a box tunnel to emerge into daylight above the yard. Then back into a box tunnel under the upper tracks, downhill to Terminal City again. The train pulls up, then backs into a yard track, the engine then heading for the roundhouse.

This 109 ft. route—almost 2 scale miles, pretty good for a home-sized layout—can of course be traveled in the opposite direction. The difference would be that a switcher would pull the train from the yard, a road engine would couple on, and move out. On returning, the engineer would leave his train on the yard lead and let the switcher shove it into the yard.

I believe that in actual day-to-day running of the layout it would prove more fun to run two trains at one time, in same or opposite directions. One train would run only over the Cut-Off; the other would run only over the Main Line. The crossovers at either side of Big Bend station would sort trains out for Cut-Off or Main-Line routing.

Built as drawn, in HO, the California Midland would require 142 ft. of flexible track, 1 RH curved switch, 3 RH switches, 1 LH #4; also 5 RH #6 switches, 6 LH #6. The layout does not lend itself to step-by-step construction. However, the Terminal City area, with yard and engine terminal could be built first, spirals next, and finally—if back of table were accessible—the cut-off. Scenery could not be built with any thoroughness until all tracks were laid and running well.

CALIFORNIA MIDLAND RR

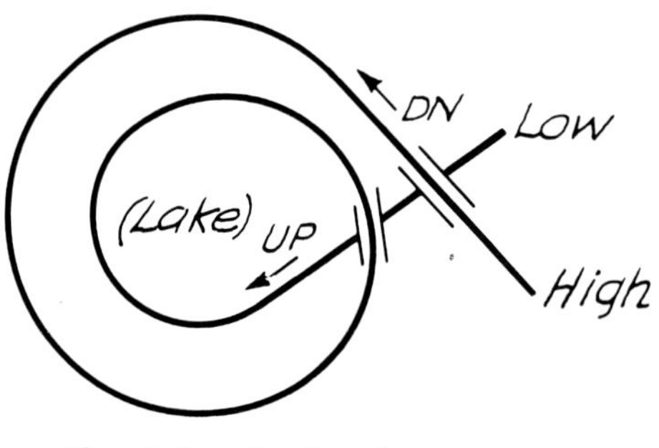

Inside Spiral

Although designed for HO, the California Midland could be built in other gauges by using these conversion factors, multiplying all dimensions by:
- for O gauge, × 2*
- for S gauge, × 3/2*
- for TT gauge, × 3/4#
- for N gauge, × 1/2#

* for O and S gauges, town inside spiral at right would probably have to be eliminated and access space put there for reaching crossover switches.

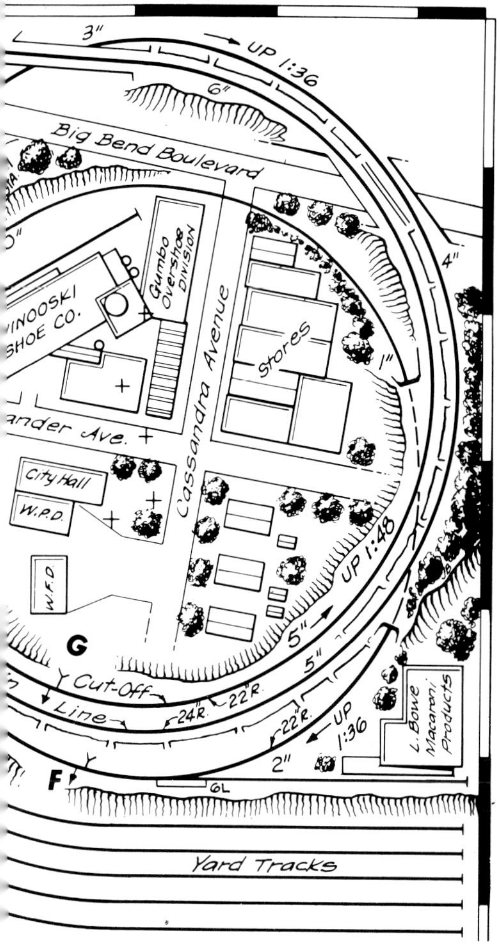

The Contributors

Al Westerfield authored many layout articles for Railroad Model Craftsman until moving to the Chicago area where he started his own model railroad manufacturing company specializing in highly detailed and highly accurate HO rolling stock sold under the Westerfield label.

Don Fiehmann is another prolific Railroad Model Craftsman author specializing in the electronics area, including such things as an HO battery powered RC diesel that rolls on the floor. Don also has a great HO railroad and has authored a book on model railroad electronics for Kalmbach Publishing Co.

The late William Schopp started writing for Railroad Model Craftsman in the 1930s during his days as a teacher. An expert watch repairman, it was only natural that his talents were used in brass locomotive kitbashing, creation of special trackwork, experimenting with HO traction and HOn2 narrow gauge, and doing RMC's Layout Doctor articles for many years.

William Livingston is a talented writer, draftsman, model builder and most of all an artist. He combines all his talents in model railroading. He's worked in several scales and gauges. His cartographs are a good technique to learn since they make it easy for a modeler to visualize a future layout.

The Timber Ridge Logging Railroad was a narrow gauge railroad serving as a feeder to the standard gauge railroad which loops the railroad mostly with hidden trackage. A lift section makes entry to the controls easy but those in smaller scales might prefer a duck under to avoid line-up problems.

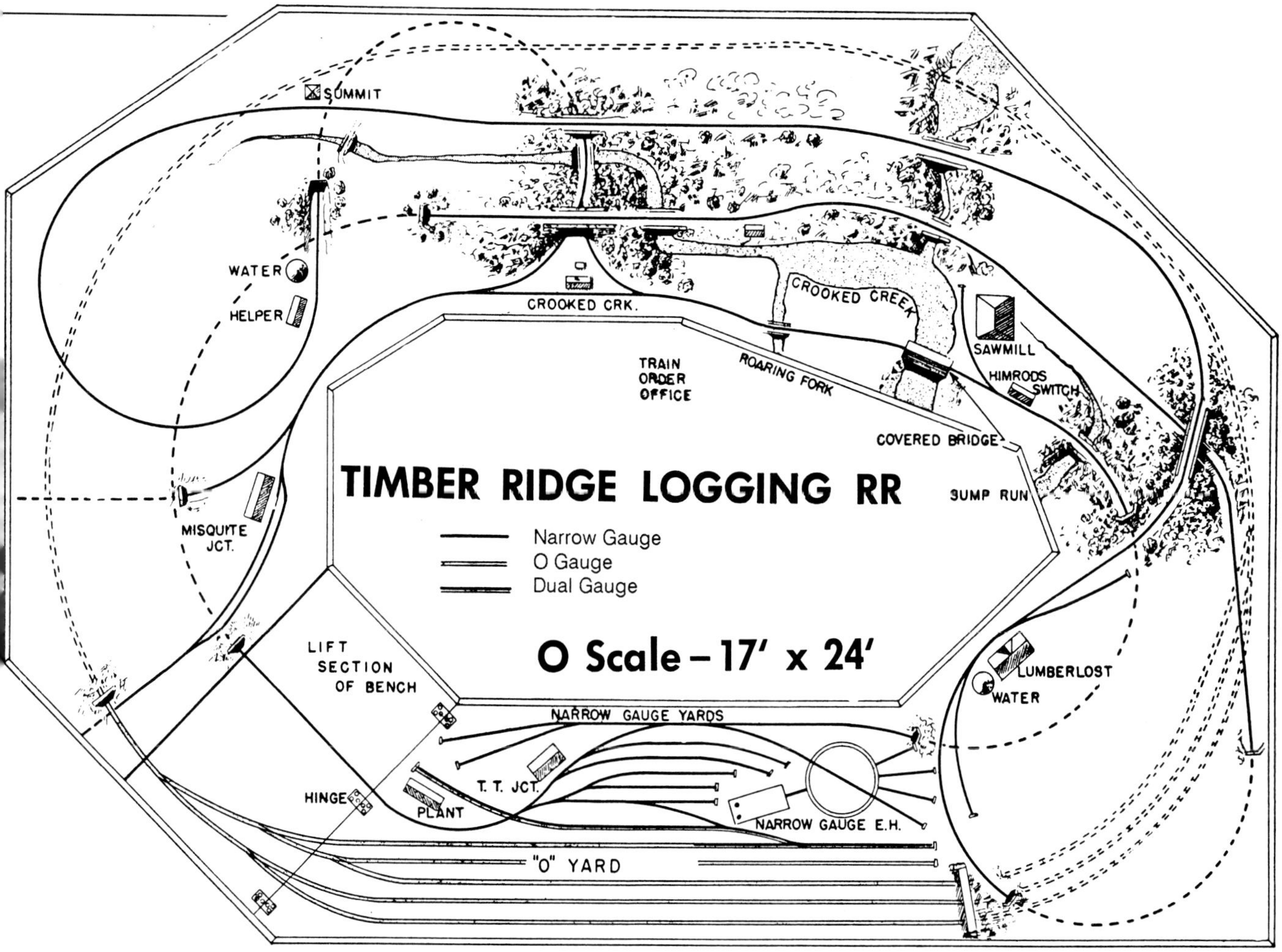

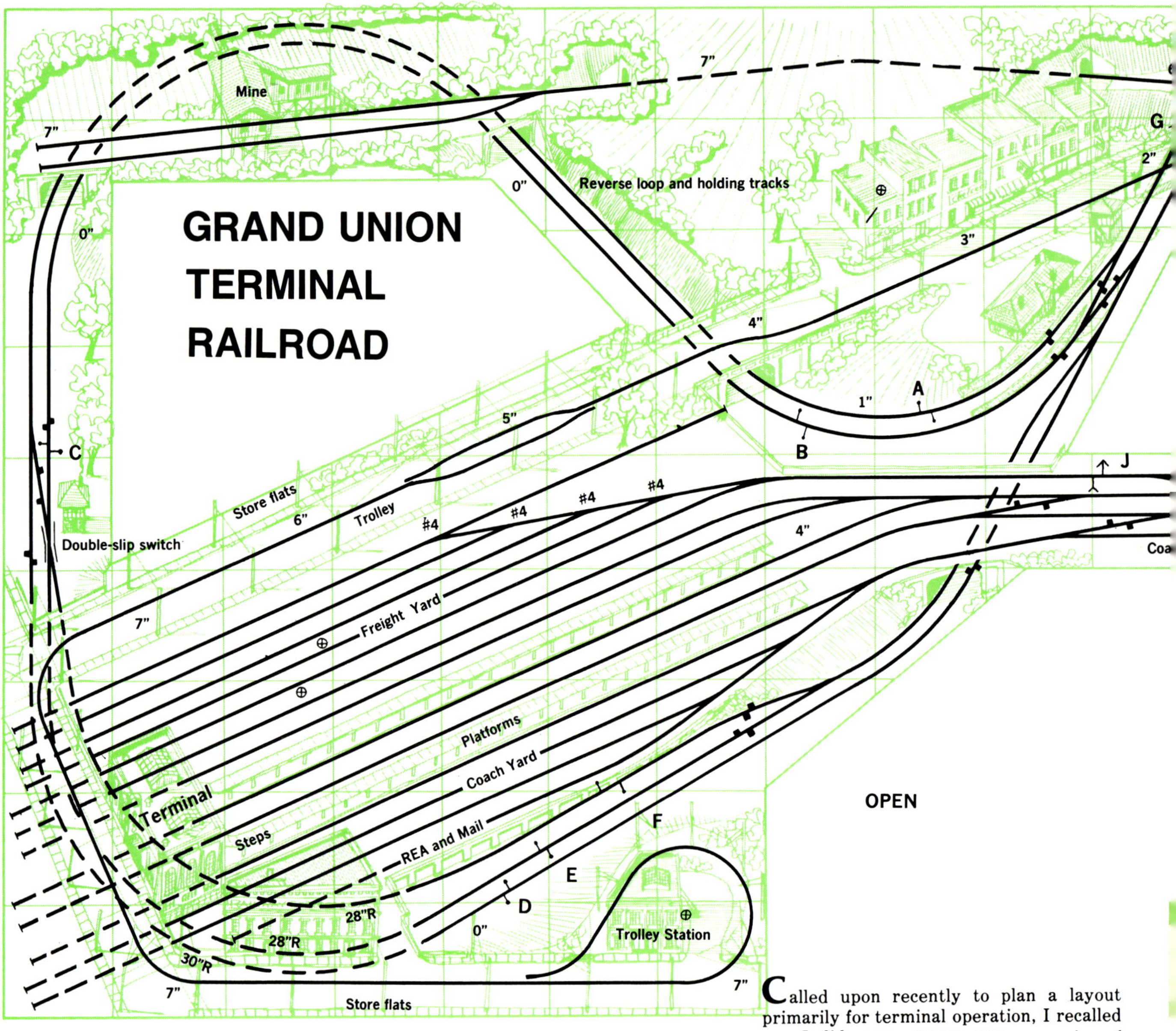

Grand Union Terminal Railroad

by the layout doctor

A trackplan designed around a large freight and passenger terminal is the basis for this month's layout suggestion. A trolley line which crosses the railroad at grade has been added for scenic and operating interest.

Layout Symbols

	Reverse Loop Feeder	0"R	Curve Radii
	Power Feeders	0"	Track Elevation
	Insulated Rail Joiners	#4	Switch No.
⊕	Major Curve Centers	Switches #6 Except Where Noted	

Called upon recently to plan a layout primarily for terminal operation, I recalled one I did many moons ago, a series of full page drawings showing various interesting aspects of passenger terminal operation running through the early part of that year. I adapted that plan, which was wider and shorter and used #4 switches, to fit the present 9 x 15 foot space in HO, and taking advantage of lessons learned in the meantime.

The earlier pike had three scissors or double-crossovers which I wanted to replace with simpler formations, even double-slipswitches (double crossovers not only require careful installation but continuous maintenance). I once warned the late Sam Skean of this, which he took as a challenge. In about a month of intermittent tinkering he did get it working to perfection, with a pair of the original giant size Challenger switch motors—he could zoom his Bowser articulated, with train, through backwards! But he later confessed that if he had it to do over again, there wouldn't be any scissors crossover.

The earlier pike also used #4 switches, whereas with the extra length there is now room for #6's except in the freight yard. I also had room for wider curves, though not as big as I really like. It is my opinion that if one has the space, the usual curve should have at least 30"R, though on the

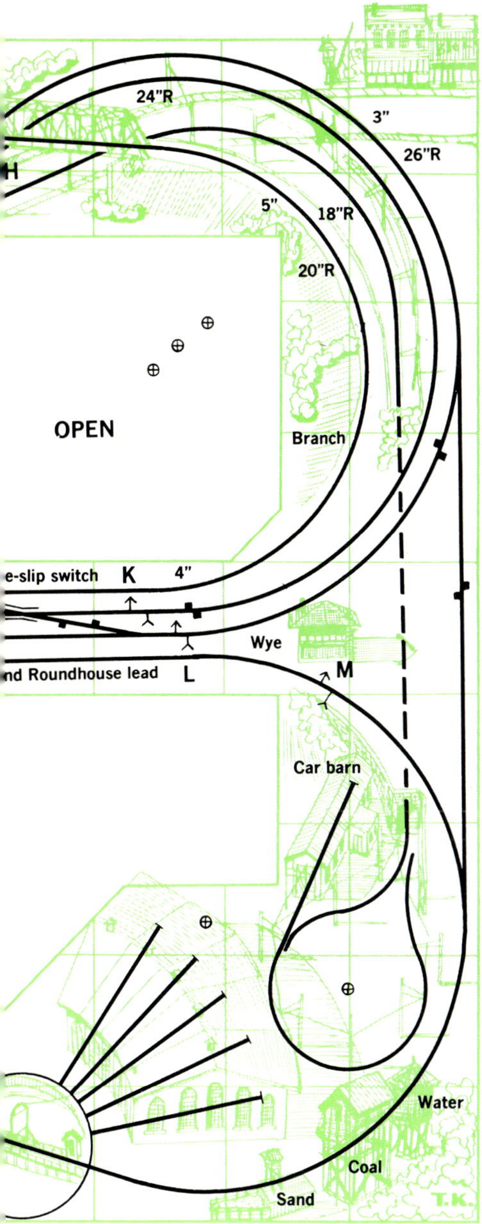

GUT RR I have dropped down to 24" and 26" for the main line with some larger, and one (to the branch, doubling as freight yard lead) smaller.

The GUT has a four track passenger terminal (two platforms) which is about right for most model pikes, plus two passenger storage tracks and one REA and mail track. The freight yard shows six tracks. Both yards have their own switching leads, the freight one turning into a branch line to a mine, as mentioned, and the coach lead turning into an access line to the turntable, roundhouse and wye. This gives a four track entrance to the terminal: the two leads just mentioned, and the two main line tracks. So a nifty terminal throat can be worked in, an interesting to look at (and to operate) feature of the line. The one shown is really the fourth version of the throat I've designed. There just is not the length to have full flexibility of running from track to track by crossovers and slipswitches, and I doubt if 100% flexibility is required once operating patterns are adopted to a simpler plan.

The whole layout is what is known in layout books as an out-and-home design, which is to say that trains leave a terminal, run out, around a reverse loop, and back into the terminal again. The reverse loop here is really several tracks. At lower left, buried under the terminal building itself, it has three tracks, and at upper left, two. Even if one track is to be kept free for "through" trains, one can still hold three other trains on the reverse loops. Direction of travel on the loops should be counter-clockwise.

If the GUT is built in O gauge, the size would be doubled; in N gauge, halved. However, I would like (as usual) to put in a plug for building the line in N gauge in the same area as HO, but with track centers and vertical clearances adjusted for the smaller size. Of course, in gauges other than HO, only the track footage would be changed while the number of switches would remain the same.

HO needs, not counting the trolley, are: one #4R, four #4L; ten #6R, nine #6L, two #6 double slipswitches; two 45° crossings, and 205 feet of flexible track.

Cabs, for control of one train over the entire railroad, could be located in each of the open areas behind the trolley passing track and on both sides of the terminal throat. I would consider the one outside, reachable without ducking under, as the main control panel.

The trolley line is something to add scenic interest, running in built-up city streets along the lower left, left side, and then across the center of the layout. At right, behind the roundhouse, it is on private right-of-way. All switches on the car line are spring switches so that the line should operate without special attention. If it is operated from overhead wire, no special electrical treatment on the reverse loops at its ends is required. I suggest separate wires for each direction of travel, as many electric lines did, close together over the center of the track. This would eliminate frogs in the wire and one prime source of trouble on model trolley lines. If an interchange track with the GUT R.R. is desired, work it in at upper right near the point where the trolley and railroad cross on a pair of 45° crossings. Such an interchange would enable railroad freight cars to be pulled with a trolley freight motor to a siding or two which could be located along the line.

The trolley line could be specially treated so that two cars could operate continuously, making positive meets at the siding. This can be done even without relays. It would also be possible, by stringing two pair of "wires" (total of four) about where I have the tracks, to operate a long model trolley-bus line.

HO – 2' x 10'
CNJ NEWARK TERMINAL

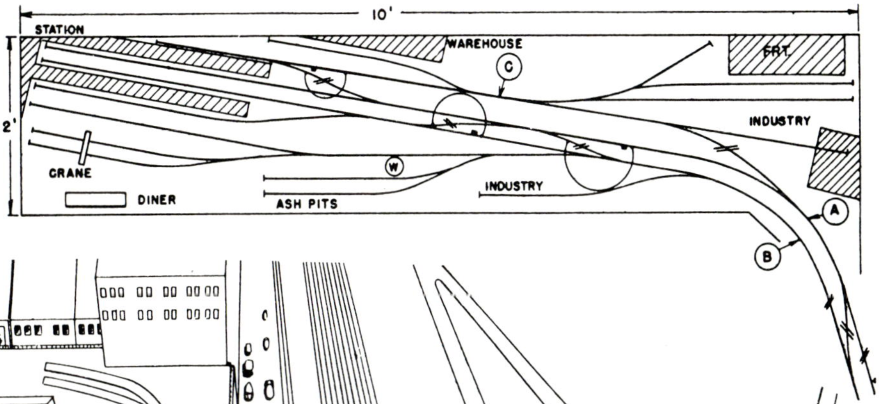

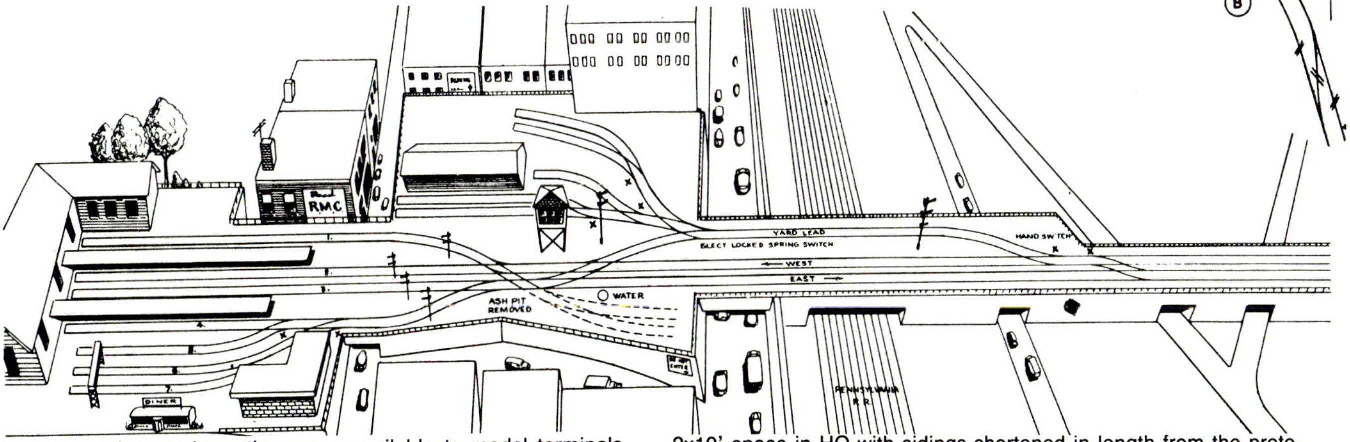

Few home layouts have the space available to model terminals such as once found in St. Louis and Chicago. Modeling such facilities often requires simplification or foreshortening to fit available space. The former CRRNJ Newark stub end terminal was designed to fit in a 2x10' space in HO with sidings shortened in length from the prototype. Few home layouts need a passenger terminal larger than this. Through terminals are a better choice for many model railroads lacking room for the stub end terminal.

SOUTHERN INDIANA RAILWAY

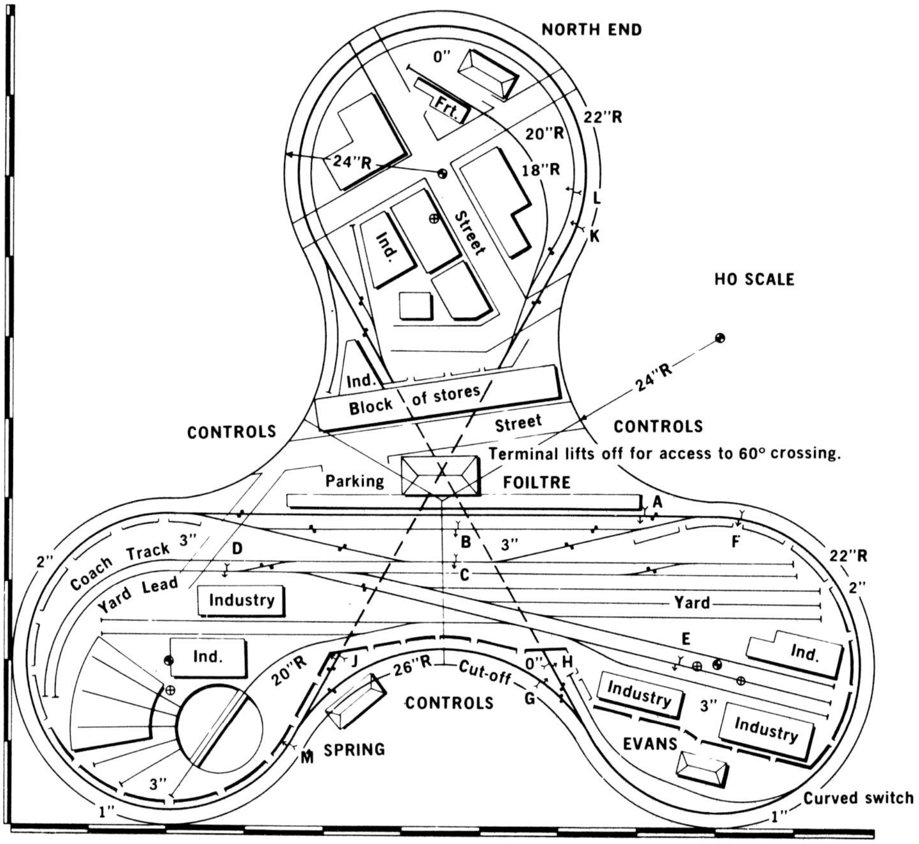

The layout doctor has some second thoughts about the possibilities of the trefoil and bases three track plans on the concept.

Trefoil Track Plans

by the layout doctor

The "trefoil" layout table was first presented in the May 1968 RMC in "Trouble Shooting" mainly because I was afraid I was putting down the greatest model railroad idea since 2-rail. As I depicted the plan, it was based on six circles tangent to a central circle, all of the same radius. It had three lobes, 120° apart, with bays between, as on the Southern Indiana plan presented here.

One feature of the trefoil, as expounded by its originator, R.W. Darby, was that the table could be made up with three identically shaped segments, for semi-portability, with a central tripod leg much like many dining-room tables, which would also come apart for portability. Some rather classy carpentry work would be required to do this, especially with the curving edge piece, but not beyond the ability of many home woodworkers.

It is now apparent that the trefoil does have better layout possibilities than I first believed, at least for certain purposes—portability, exhibition, etc.—though I still say not as good as an around-the-walls layout in a room large enough for it to stand free. The first plan presented here, the Southern Indiana Ry., is based on a very similar plan sent me by John Lord. Here R=24", so that each lobe of the trefoil could be made from a standard 4' wide piece of Homasote or plywood. Maximum curves, following the edge, would be 22" radius, though following a concave edge gives a 26"R.

The Southern Indiana layout consists first of an "oval"—rather a water-wings—on two of the lobes rising up from 0" at its lowest level to 3" just in front of the terminal station at Foiltre. The entire yard, engine terminal, and industrial area, which fills the center of this oval, is at 3" elevation. The grades up from the lower regions are rather steep, 1:36 or just under 3%. Note that the grades do not stop before the switches at Foiltre, but continue through the initial nests of switches at each side and end of the terminal area; *then* the grade levels off.

I've revised Lord's original only slightly, chiefly with a few more industrial sidings, and a change from a parallel track enginehouse to a turntable and roundhouse. I've drawn it for top-mounted switch machines and insulated frog switches. These can be either #4 Casadio with a radius of around 18" and an angle of 14°, or #4 (really 4-1/2) Custom-Line with a radius around 22" and an angle of 12-1/2°. However the right-curved (RC) switch at Evans would have to be Casadio since they alone make curved switches of this type.

From the lower regions of the front two lobes (Evans and Spring) tracks swing off, cross under the terminal, and loop around the back lobe where there is the town of North End and a passing track.

I've divided this pike up into twelve control sections lettered A through M, excepting I, each isolated by twin insulated rail joiners at each end. They may be all controlled together, wired in parallel, but I think the best system would be to use a bank of 3 Atlas Selectors and two power packs to give cab control for two trains at one time. In this case all the arrow-tails would be wired together and then to the power packs per the Atlas wiring book.

Mr. Darby, the originator of the layout shape, came through with a very interesting N gauge layout on which my Big Three System plan is based. I did detect one thing of interest; he "cheated" on the basic design of six circles surrounding a main circle, all the same radius. His central circle is larger, here 26" radius, with 18" radius for the lobes and bays. This is shown in the northeast part of the Trefoil Traction Co. (HO) plan. It gives sides to each lobe which can be straight and almost parallel. This makes a slightly more practical site for a layout, though the design system then is subject to some variation. Actually, I "cheated" or misunderstood his basic system, which was not quite the seven circles I used.

My Big Three System is based on Darby's design, but with "improvements" which I feel a Layout Doctor is entitled to make. My main change was to make all trackage one-way, something I like to do since it simplified controls and makes automatic signaling (if you ever get to it) easier. He had two crossovers between the double tracks west of Flying Junction, and another pair in the area of Trefoil. He also had a double tracked, one level junction where I have a Flying Junction at the town of the same name.

The route of the Big Three is really a double tracked main line connecting reverse loops at Trefoil and Valley. However, addition of cut-offs I and II between Flying Junction and Valley, and between Flying Junction and Montana Jct., gives two variations on the route. The tunneled parts of both cut-offs could be arranged to be holding tracks for trains not being run and supposedly off at the other end of the giant system.

Even the intern Layout Doctor would notice that the main track, downhill from Trefoil to Flying Jct., also has to be used as a switching lead for the four-track yard. True enough, but this could be easily overcome by putting in a separate concentric yard lead, with three tracks down the hill, using the inner curve for a yard lead. Actually I think the better thing to do would be to install a left-end ladder so that trains could enter the yard at the left end. This is

BIG THREE SYSTEM

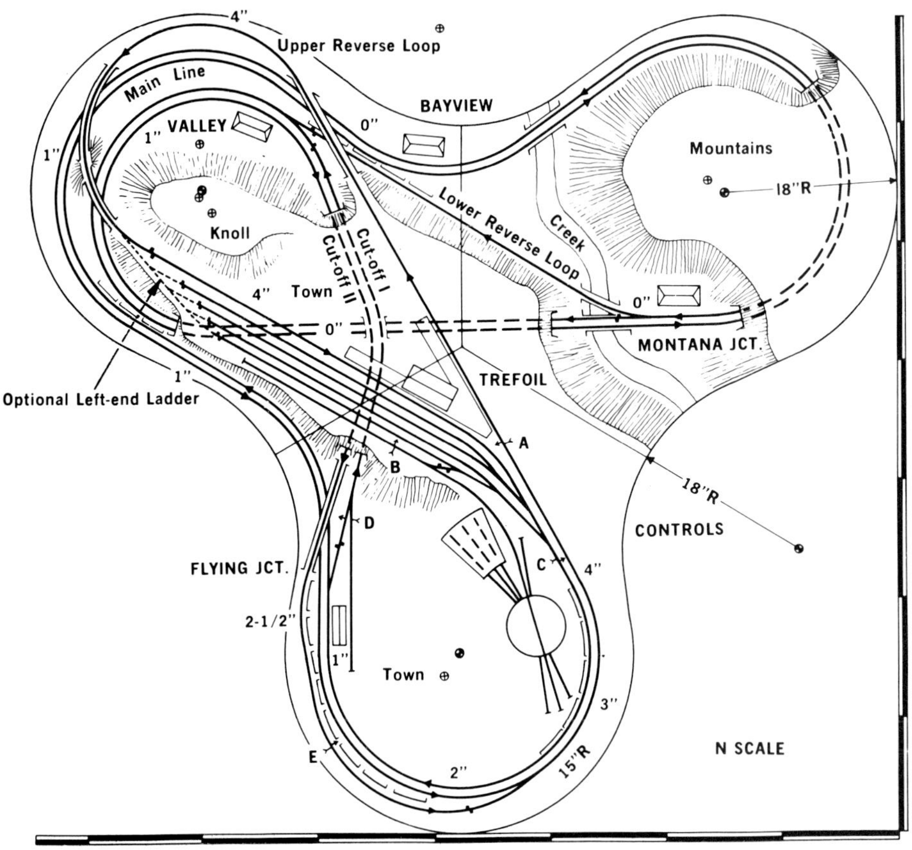

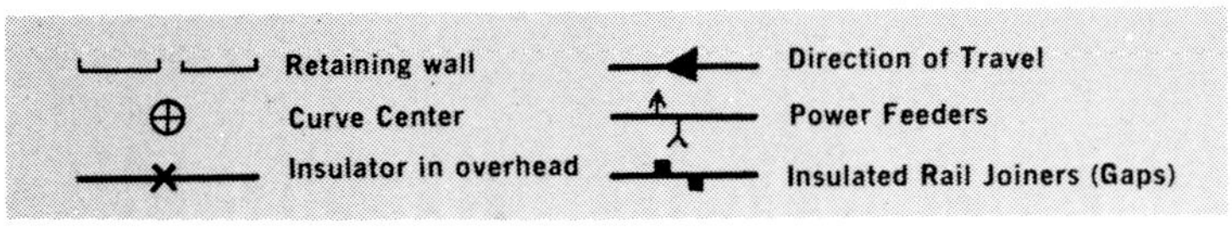

TREFOIL TRACTION COMPANY

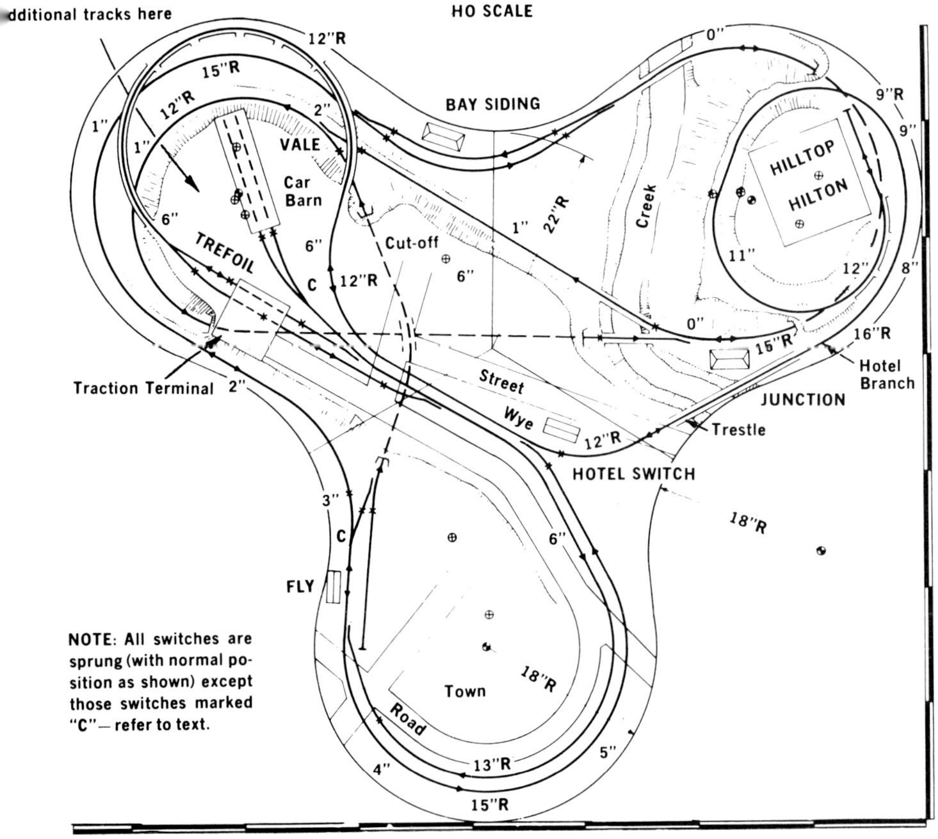

shown dotted. I believe that on a layout of this size, running would prove more important than niceties of switching, and trains would be more or less be kept made up, ready to roll.

Electrically I show each of the double tracks and half of each reverse loop as a separate control section—A and C. In reality I think that you would want to subdivide each into shorter blocks, possibly manual, possibly automatic blocks with signals. The enginehouse lead (B) and the two cut-offs also form separate blocks.

I took the very same table as Mr. Darby and redid the Big Three into a Trefoil Traction Co., HO gauge trolley, mainly single track. After it leaves its principle town with a traction terminal and car barn (note suggested location of additional tracks to the side of the barn), it goes downhill, double track, to Fly where there is still a junction. However, the main track proceeds around the left lobe and right lobe of the trefoil, with a passing track at Bay Siding. At Junction, the lower reverse loop splits off and swings around to return a car over the same route.

The cut-off from Fly to Vale short-cuts the outbound trip and starts a car on its return to Trefoil sooner. Again the tunnel could be used for hiding a car for a while. Please note that the controlled switch (C) at Fly is the only switch on the whole pike, outside of the terminal area, which has to be controlled; all the other switches can be sprung (as drawn), a great short-cut in model, even as in real trolley operation.

For a little added interest I've added a local trolley from Trefoil up to the Hilltop Hilton hotel. This branches from the main at Hotel Switch, and surely a Birney could meet all main line cars there. However, for added interest I suggest that the inbound main line from Hotel Switch to Trefoil terminal be made two-directional so that the local car can use that both ways, ending up on the left end of either of the through tracks at the terminal. People would then have a cozier place to wait for the trip up to the scenic inn.

No wire feeders are shown. However, I show X's in the line to suggest where I would put insulators in the overhead wire. Each section thus created would then have its own feeder. With overhead wire operation, as with third rail, you do not have to worry about special wiring for reverse loops. And with trolleys running overhead, you can wire them so that they are always going forward when the rear pole is up (on any given polarity of track and wire, that is).

No tabulation of the number of switches, etc., is shown for these three pikes, though you can easily count them, adding any new ones you might want, subtracting those you want to delete. If you want to see how many feet of track are required, open your dividers to one foot as indicated by the rules around the edges of the plans, and step off feet-of-track, omitting switches. Be quite generous to avoid having to complete your trackage by combing a lot of offcuts.

O.K., it's possible to have interesting track plans on a trefoil. Whether anyone but Mr. Darby actually builds one remains to be seen. If anyone does, RMC would like to see photos and maybe a story on the pike.

Combine a Railroad with Canals

A canal with locks and a bridge over a river is the scenic focal point of this twice-around oval trackplan.

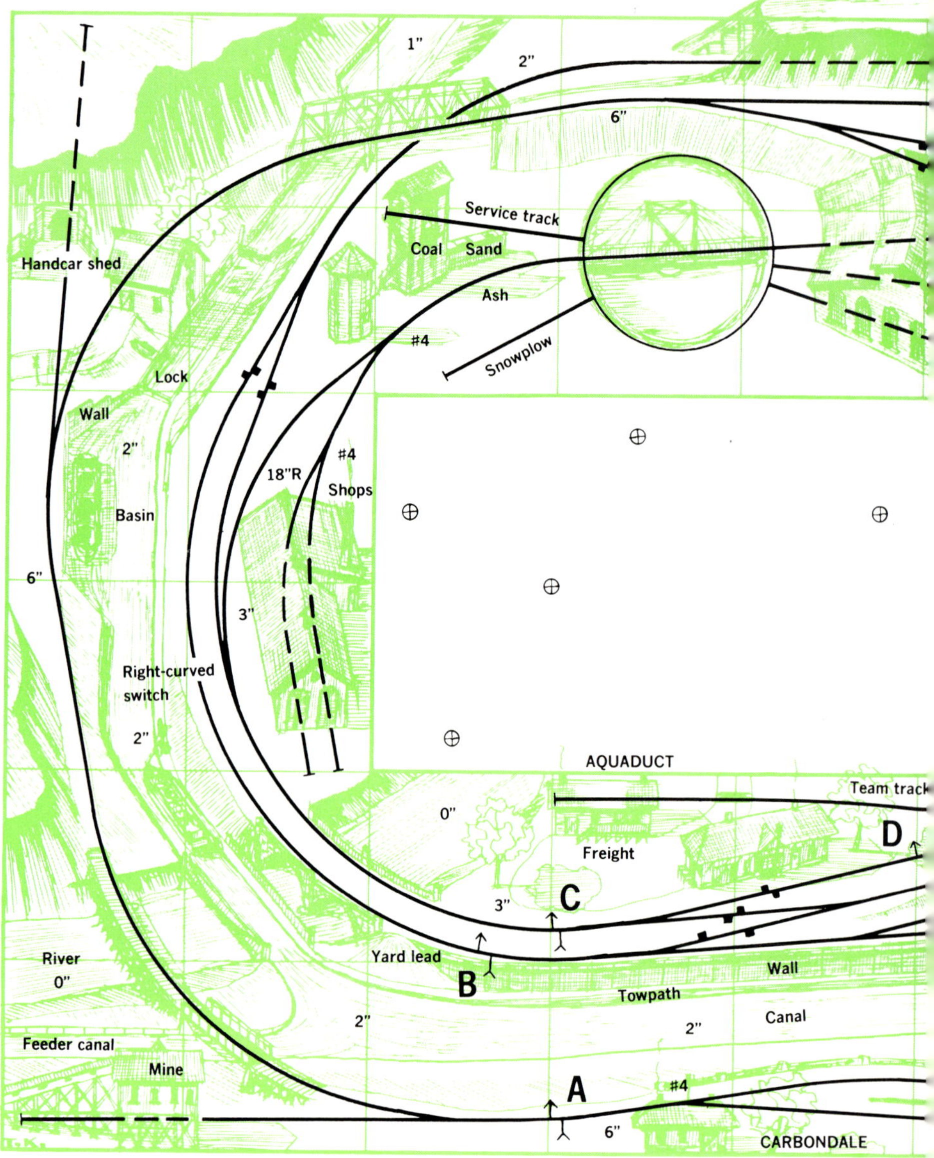

I came across a layout drawing I had in RMC called Black Diamond R.R. & Navigation Company which attracted my attention. (Coal was commonly called black diamonds on account of a similar composition: hence the Reading's diamond-shaped herald and the Lehigh Valley's once popular name passenger train — by the way, "Reading" is pronounced "redding", not "reading".) Here was a combination railroad and (dummy) canal layout with curves of 36" and 34" radius, which I liked so much that I thought of asking the editor to reprint it some day, since it is by no means obsolete. However, I decided to redraw the plan with smaller curves — 24" and 22" basic radii — and in a smaller space — 6 x 8 feet instead of original 8 x 10. It suffers little in the reduction, radii reduction compensating for reduction in scenic acreage. Besides, the curved switches now on the market permit a few improvements in siding location.

The United Maryland R.R. & Canal Co. is the result, so named as though the B&O *Railroad* and the C&O *Canal* were "married" as the Camden & Amboy R.R. and Delaware & Raritan Canal Co. once were in New Jersey. The layout plan itself is an inside-out figure-eight (also called twice-around oval) with the non-working canal separating the two lobes of the 8 at left and bottom. It might be called an "inverted" layout, since instead of having the lowest tracks at the front and higher tracks back a little, it has the higher track along the front (bottom of plan) with lower ones inside, reached (in HO) from the central access pit. This, by the way, is one was to "kid-proof" a layout, at least while the youngster is just starting to walk: remove Carbondale station, the mine buildings, and any cars along the front, and the railroad is safe from little hands — at least until he learns to crawl into the center of the layout or climb up on the table and do his Gulliver act.

I show the UMRR&C with unit switches having integral switch motors, which in HO means either Custom-Line or Casadio or both. For a railroad operating during the latter part of the canal era, this is, of course, anachronistic. A late nineteenth century layout would have hand-thrown switches, or at best ones worked by rodding from a signal tower. With either Custom-Line or Casadio switches, code 100 rail would have to be used, probably in flexible track, again anachronistic. It would be better to use code 70 rail, possibly on individual ties (use Kadee's Rail Spiker in HO to save much time).

I'd recommend building such a layout on an open benchwork table since nearly all track is above "sea-level or O". Begin construction at O" (the ends of the yard sidings under the hill at upper right) and proceed clockwise through Aquaduct (along the top), bridging the original yard tracks, and up past Arox and Carbondale. You could then work counterclockwise from Aquaduct and drive the Golden Spike somewhere leftward of Carbondale. Since the canal is really scenery, that could be added along with other scenic work. If real water and locks are desired go right ahead and try it. I once concocted a layout plan with an oval shaped working canal (I didn't explain how it would work) with connection of the high and low ends of the canal by means of an inclined plane under mountains or behind the scenes.

Trackage required for HO would be: seven #6 RH switches; three A #6 LH; One #4 RH; four #4 LH; two RH curved; and 90 feet of flexible track. For N gauge, same sized table, these would remain the same — if table size were halved, flexible tract footage as well as elevations would also be halved. In this instance, the central access pit would drop out, being an impractical one foot wide, and the table would be solid. For O gauge, all linear measurements would have to be doubled, but numbers and sizes of switches would remain the same. For S gauge, lineal measurements would be 1-1/2 x HO; for TT gauge 3/4 x HO, or to be more exact 24/37 x HO which is closer to 2/3.

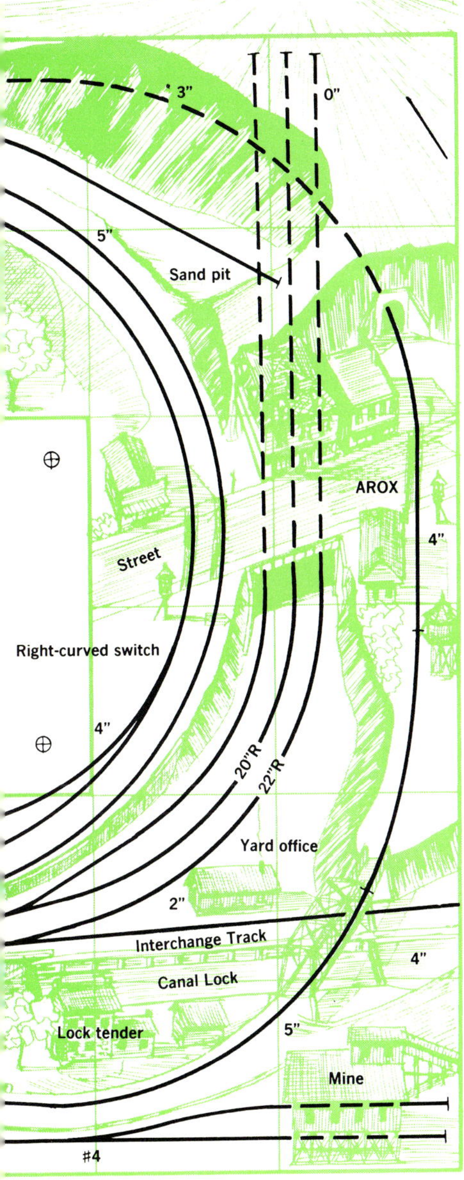

The Utah Railway

Many modelers prefer to put as much trackage as possible into the minimal space available to them. Try this spaghetti bowl layout which packs a long run into a 4x8 area in HO, about 3x5 in N.

A "bowl of spaghetti" is a term of derision flung at model railroad layouts that are all trackage. However, one school of thought holds that, if you're going to model railroad, why not *railroad,* making your layout the maximum that you can build in the relatively small area, with due consideration to good appearance and your ability and time to keep up with maintenance?

The 4x8 foot (in HO scale) Utah Railway is an example of a spaghetti bowl which should be within the capabilities of a model rail of only a bare minimum of experience. You could begin laying track on the level table, starting at the #4 Custom-Line left-hand crossover at the lower right. To its left, it enters a double-track tunnel of 18" and 16" radius curves; the inner one forms a reversing loop around the pond and yard, while the outer one rises in the tunnel to emerge at a higher level.

To the right, level curves of 22" and 19-1/2" radius are concentric, then merge together with a turnout (switch) at the rear, near elevation 2". These have wider than usual spacing since the outer is a main line, the inner a switching lead, and friction between cars on these lines must be avoided. The yard and enginehouse can be easily built at this time since they are

Layout Symbols

Symbol	Meaning
	Power feeders
	Reverse loop feeders
	Insulated rail joiners (gaps)
1:36	Grades
A	Electrical blocks
⊕	Major curve centers
0"	Track elevation

Size of squares: 12" for HO scale (6'x8')

Grades 1:36

Switches No. 6 unless noted

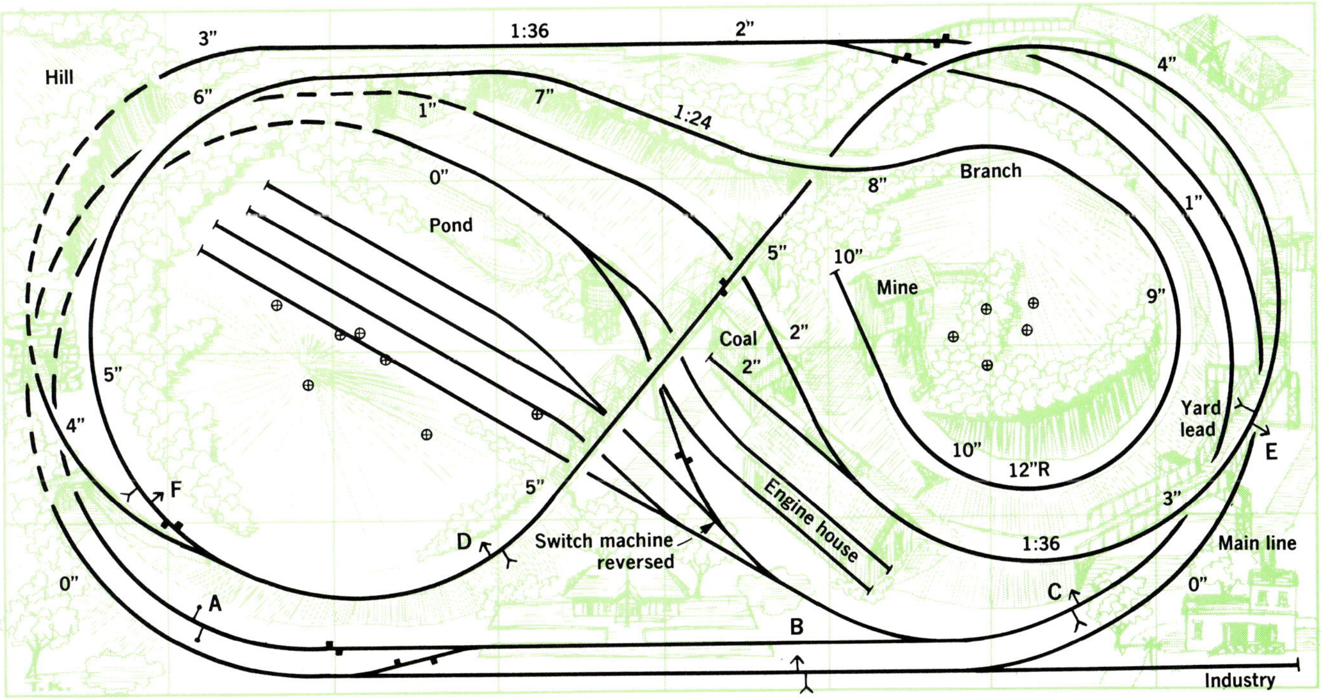

Modern diesel era railroads require fewer yards, engine terminals, multiple track mainlines than do steam age railroads which need coaling towers, icing platforms, water tanks, passenger and freight stations, turntables, ash pits and other special facilities not seen on today's railroads. The Susquehanna Northern operates back about 1939.

all on the level. Note that one yard switch by the enginehouse has its motor flipped (assuming sectional track turnouts are used) to the curved side to permit close spacing of switches.

To build the upper levels, beyond the 1-1/4" and 2" elevations mentioned, I'd suggest starting at left and build forward behind the pond, since this track will go under the long bridge. This line, when it gets to right front and right rear, has two curved trestles or other bridges. It is difficult to design and build these from the standpoint of a bridge engineer, so just do the best you can to maintain the curves and clearances and at least a semblance of realistic appearance. If a critic tries to shoot you down, ask him for his best suggestion for better bridges (or ask him to build them for you).

If you start laying track at right rear at this time (at 2" elevation), you can curl around to the left end so that only the lack of a bridge prevents a continuous run connection with the line which curls to the right rear (with the two curved bridges). It is then your task to build that high bridge. With a 5" clearance (3" at extreme right), use deck girder bridges if you like. The location of the piers may vary somewhat. If the right-hand Snap-Switch near feeders F had a right-hand curved switch substituted for it, probably the whole bridge could be moved back a few inches and/or its angle changed, so let the bridge position work itself out as described.

The branch line, which has sharper curves and steeper grades than the rest of the road, is for shoving cars up to a mine and hauling them down again. It is a separate electrical control block (F) and would be ideal for a geared engine or even electrification.

Additional control sections are: A, the reverse loop (there is no "re-reverse loop" mate, since the present one is mainly for turning engines and hiding short trains); B, the main line; C, the yard lead and yard, excepting the reverse loop; and D and E, the 3% grades leading up to the long bridge (it is better to keep them separate since full power may be needed to climb the hill, very little to descend). Switches, excepting the left-hand crossover at lower left and the yard lead switch at upper right (which are Custom-Line #4's), are Snap-Switches. Otherwise, trackage is all flexible since the variety of radii makes building this particular pike with sectional track components impossible.

The limitless 4x8

by Al Westerfield

Artwork: the author

The author offers a new look at an age-old basis for many model railroads—the 4x8 sheet of plywood—by analyzing the use of viewblocks to create many separate scenes.

It never seems to fail: No matter how large a space a model railroader has available for his layout, it is never quite big enough to create that one great scene he's been longing to build. The guy with a 4x8 board dreams of a small room all his own. And the guy with a room wants a full cellar. But, worst of all, the guy with a cellar has visions of a quonset hut. Yet, through all these evolutionary changes (the serious modeler always finds some way to expand), the dream remains unfulfilled. The answer certainly isn't invariably lack of space.

What most persons don't realize is that almost everyone has room for at least one or two great scenes. They only require planning and execution. The execution is up to the individual modeler, of course, but I contend that if a scene has enough potential, the modeler will rise to the occasion and do better work than he thought himself capable of.

Planning requires concept and scope. The concept of "Ohio short line, circa 1955, late steam and early diesel" is not enough. It must be refined to include your modeling likes and dislikes, abilities and shortcomings. For example, if you hate to lay track, stay away from double-tracked mains. The scope deals with how small the detail and how encompassing the scenery. The scope with which a layout is developed will definitely limit the number of concepts you can use. For example, the layout at the B&O Transportation Museum in Baltimore, Md., is a breathtaking display of main line trackage through the mountains. It hardly lends itself to peddler freights and industrial switching. The Gorre & Daphetid, with its spectacular vistas, still had plenty of work for the peddler (local) to do. Only rarely can both concepts be included at the same time; the scope of each is simply too grand. Most people compromise to one extent or another. Allen McClelland's Virginian & Ohio is an excellent example of this as he has skillfully synthesized most of the facets of the prototype into a manageable area.

The concept of a layout is rarely visualized as a whole, however, except perhaps at the moment a visitor enters the room. The eye is invariable drawn to specifics, be it a scene or an individual model. As a result, it is not often important that the concept and scope be uniform. What is important is that the transitions from one scene to another be smooth. Nothing is quite so jarring as the unintentional incongruity—the big city yard in the middle of the Rocky Mountains, for example. One way to avoid this sort of thing is to duplicate nature. But where nature takes hundreds of miles to accomplish these changes, we have a few feet. Another way is to take advantage of our penchant for specifics and build several scenes of limited scope in such a way that only one can be seen at a time. The concept then becomes one of creating several individual and perhaps vastly different scenes to suit our tastes.

I've selected the 4x8 area (the size of a standard sheet of plywood) as it is the most usual starting point for neophyte model railroaders. Fig. 1 shows some of the ways this area can be divided up. The lines represent scenery dividers that are high enough to prevent seeing more than one scene simultaneously under normal

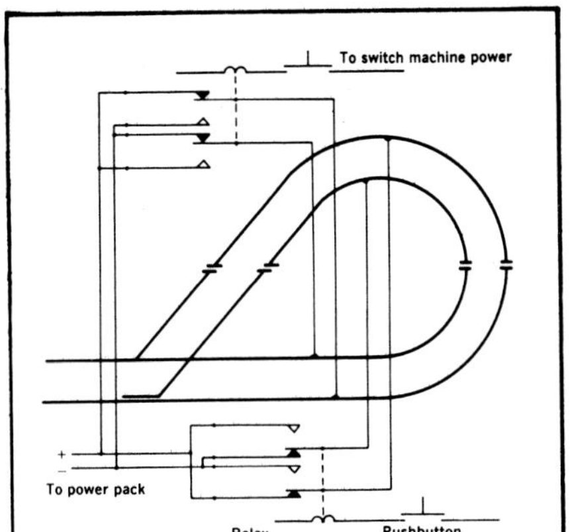

Fig.2

viewing conditions. Plan A is a simple bisection giving a ping-pong-table effect. This immediately takes away the greatest visual effect a model railroad can offer, the long run. Plan B on the other hand, uses the space much more effectively. Notice that it is a natural for modular construction. Two 2x8 sections can go places a 4x8 cannot. Trackplan I shows some of the potentialities of this plan. I've included plan C simply to remind you that the dividers need not be straight. Diagonals, as in D, offer some interesting possibilities, such as a double-yard layout as shown in Trackplan II. Small areas are naturals for dioramas as in E. Plan F provides a slightly larger area for the diorama without interrupting the "long run." Trackplan III shows how you can turn this into a real knockout. Our final bi-section is G, a plan that yields an even larger diorama. The L-shape might be suitable for a crowded city scene.

A general rule seems to be that in cutting up a layout into two or more scenes, the difficulty of preparing a good trackplan increases with the number of scenes. Our first tri-section, plan H, is a good example of this. The center section is almost claustrophobic. The only possibility would be to use it as a valley or chasm that would be spanned by several long bridges. Plan I, on the other hand, provides two small dioramas and a center section with operating possibilities. Notice that diagonal viewing is retained. Trackplan IV shows one way to utilize this space. Plan J takes our first successful bi-section and puts a diorama at the end of it. After reading about Trackplan I, imagine a diorama at the curved portion of the track and you'll see how effective J can be. Plan K tries some diagonals and creates three potential "scenic" dioramas. A slight variation yields plan L, which I have expanded into Trackplan V.

Quad-sections are quite difficult to master. Plans M and N have little to offer. Plan O has several possibilities, however. A circle of track and four prototype scenes would make an excellent display layout. Or, more positively, a novel switching layout such as Trackplan VI could be designed. The dotted line in plan O is there to remind you that there is no need for the dioramas to be symmetrical.

I have not included any further sectioning simply because they look too difficult for the small area. I'll leave it to you, if you're so inclined. But let's get on to the issue at hand: seeing what kind of layouts can be designed in the spaces we have created.

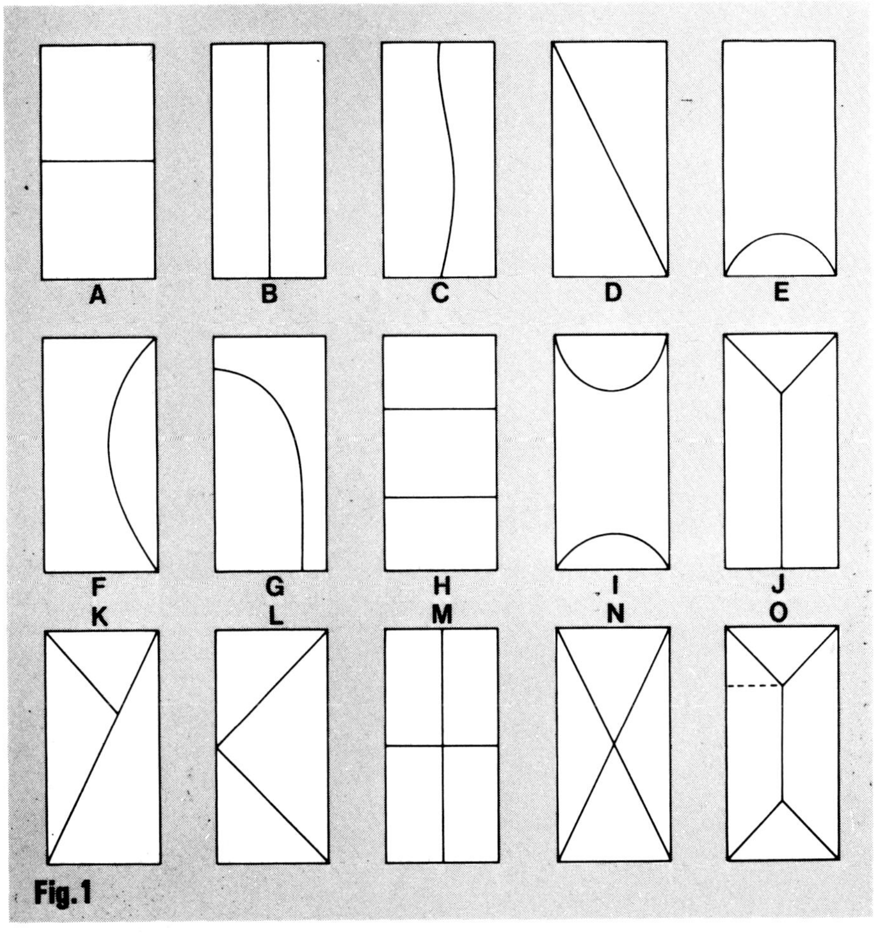

Fig.1

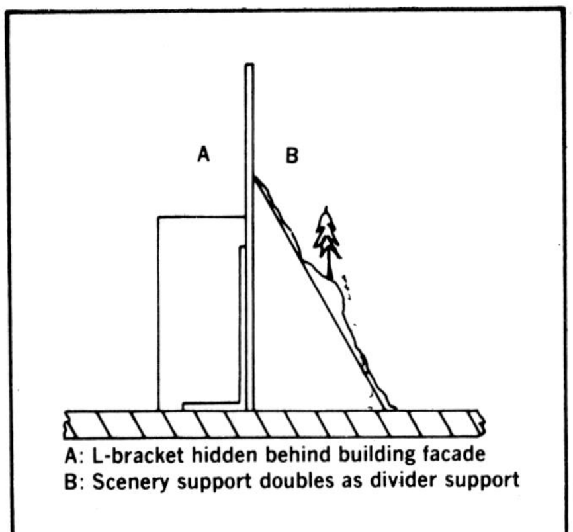

Fig.3

Trackplan I provides two long, narrow areas ideal for switching. The trackwork is exceedingly crowded, and would *look* crowded if it were not for the scenic divider. The plan as shown is essentially flat, so that a plywood base could be used. However, I would suggest that in actual practice the tracks be made to vary in height somewhat to avoid the flat look; an inch is all that's needed. The plan provides nine sidings, a passing track and a loop. The loop is a neat trick here because it cannot be spotted by visitors; usually they stand out like sore thumbs. For descriptive purposes, I've divided the plan into Modules A and B. This is not simply as a convenience, but also to keep in your mind the potential for modular construction.

Module A is conceived of as something akin to the Slim Gauge Guild's diorama, a small town at the foothills of a mountain range or tucked away in a valley. Rockwork is built up in the background to six inches or more and blended into the divider. A dock provides a source of traffic, as does the "junction" of the railroad. The mainline enters Module A via a tunnel and curves into town. Here it can branch off into another tunnel (the loop) or continue through town. A passing siding allows all industries to be serviced by a single loco. The main beyond the passing siding is long enough for at least a loco and one car, as are all dead ends in this plan. Industries are laid out to allow maximum area and avoid the "straight line look". These industries would preferably be raw material sources such as grain elevators, feed lots, etc., that could serve as sources of supply for a manufacturing center such as Module B is envisioned.

Module B gets a lot of switching in simply by cutting the area necessary for structures. The divider is lined with facades of warehouses and manufacturing plants. These serve as industries requiring trackage without taking up more than an inch or two of width. The tail of the loop masquerades as one of these sidings and is in fact long enough to serve as both. A little craftiness in your choice of buildings should hide the connection between the modules. The mainline is fairly long and unobstructed, to allow making up of consists by hand-power prior to an operating session.

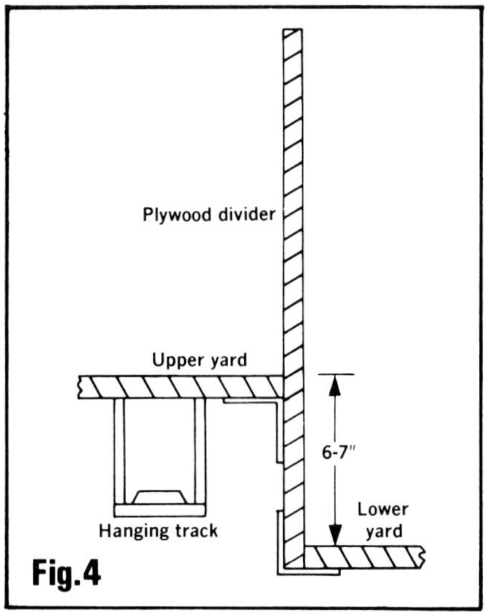

Fig.4

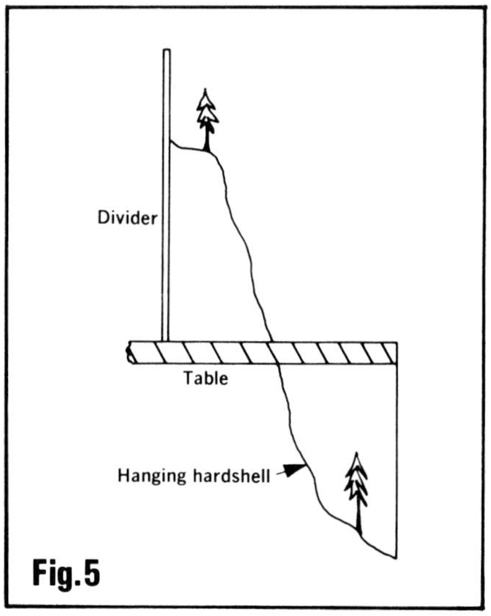

Fig.5

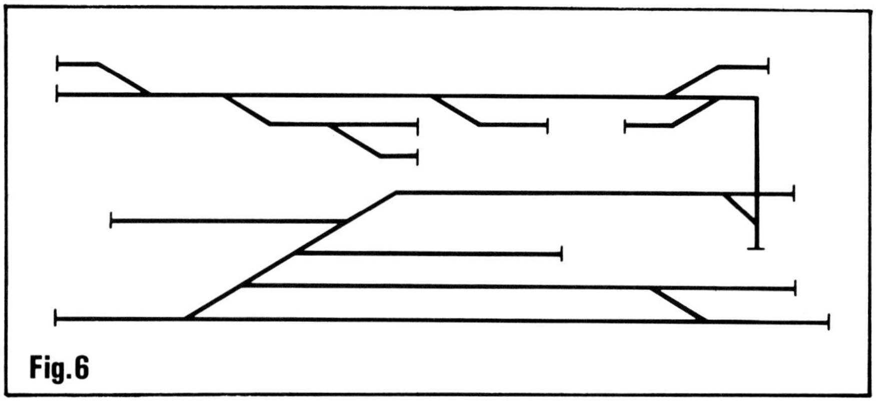
Fig.6

Electrically, walk-around control is almost a must. The traffic flow is designed with this in mind. The main transition between modules is via the long curve. It is natural to walk around with the trains. However, some special wiring is necessary since the layout is essentially a loop. Both rails should be gapped exactly where each track crosses from on module to another. This forms two isolated blocks, corresponding to each module, as shown in Fig. 2. The blocks are wired so that the curve route is always lined up electrically. The hidden connection between modules would be used principally for changes of locomotive direction and run-arounds, not normal traffic. It is situated in the center of the table. We would logically expect that the operator would send the loco through the divider, stop it immediately, then walk around and start up again. The circuit shown here follows this logic. A single pole, single throw momentary (push button) attached to a relay with two sets of make-make points is used to change the polarity of the *unseen* module for the few seconds that the loco changes blocks. Using a momentary switch prevents the major route from being electrically fouled because of forgetfulness. The change-over process would then be the following: (1) engine runs into approaching siding, Module B; (2) loop route is selected (turnout thrown); (3) power pack direction is reversed, engine proceeds onto route; (4) momentary switch is pushed down; (5) engine goes through divider, is brought to stop; (6) momentary switch is released, operator walks around to module A; (7) power pack direction is reversed (remember we are now traveling in the opposite direction!) and loco proceeds. The system is almost fool-proof. One last point: Since walk-around control is being used you can place the turnout ground throws near their respective turnouts.

Almost any stiff, smooth material may be used for the divider. For most applications, I suggest Masonite because it has some "give" to it. This makes it possible to bend it somewhat for curved dividers. A coat of white latex house paint will seal it and yield a light colored surface for scenic painting. Fig. 3 shows a variety of ways to support the divider.

We've seen that a diagonal bi-section turns the rectangle into two wedge-shaped areas. These are naturals for yards. As a matter of fact, a yard throat generally wastes a lot of space simply because nothing fits too well around it. Here, the area and track are the same shape. Trackplan II is a double yard. It is designed for that special nut we all know—the guy who always hogs the yard panel on the operating nights at the club. This plan allows him to hog *two* yards the other six nights of the week. (Who knows? Maybe saturation will set in and someone else will get the club yard once in a while.) Seriously, for those who especially enjoy making up trains, this plan supplies about all the action possible in a 4x8 area. Each yard is capable of handling 35 to 45 standard cars. The arrival/departure track has room for nine 40-foot cars, a good train length for 22" radii curves and 3.6% grades. Since I'd wager that most yard operators use plastic diesels, the curves and grades should present little difficulty. In any case, I've provided for turning motive power in one yard, so the steam fans aren't left out in the cold.

The reason this trackplan works is that yard A is 7" higher than yard B. This allows the main line to snake down and around, resulting in about 16 feet of track between yards. The grade is kept steady so that it is exactly half-way between the yards in height, at the same time it is half-way in distance. This allows sufficient clearance over yard B (although through bridges will probably be necessary) and clearance for the main when it goes under yard A. The benchwork is a little unusual, in that the divider must be an integral part of the table. See Fig. 4. Three-quarter-inch plywood is recommended for this. It must be about 6½-7" high for the table separation, plus an additional foot or more for the backdrop. The table itself is simply a sheet of plywood cut diagonally.

This layout is best operated by two stationary control panels. The long main line of constant grade gives the operator a chance to switch panels on the fly. If desired, a flip-flop control could be wired in. Of course, the layout is a natural for two-man operation, especially if a little friendly competition is your meat.

Trackplan III is a basic circle layout with a single exception: It affords the chance for some really great scenic effects. The plan and Fig. 5 give some idea of what can be accomplished. If the board is up on legs or on a sawhorse, the diorama can be cut out. Then, using hardshell techniques such as I described in the January 1974 RMC, the scenery is hung from the board forming a half-bowl. With a foot or so of scenery beneath the table's edge, another foot above and six feet of trackage running through it, a scene to rival any in the country is available. I've suggested a duplication of the trestle in the plan, but there is no way I can show you exactly what kind of scene to model. If you haven't always wanted to build a giant mountain scene and don't already know exactly what you'd do if you had the space for it, I'm wasting your time.

The rest of the layout, within its limitation, tries to use the divider to best advantage. The small yard is built along it, as are the two large industries. The yard extends to the table edge to allow easy placement of cars on the tracks. Although not designed primarily as a switching layout, the switchback to the industries off

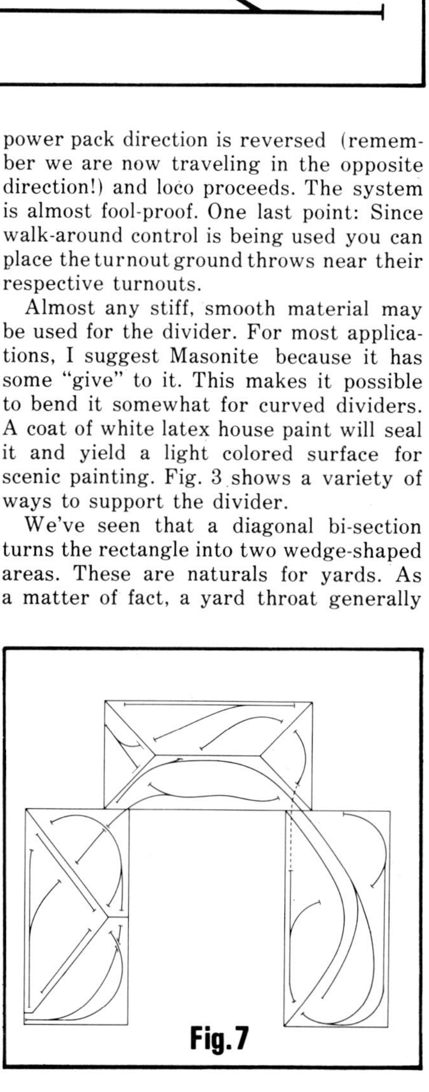
Fig.7

the yard can provide some fun. A loop allows turning of motive power, but more importantly lets the trains enter and leave the yard without backing. Several other industries are provided along the main to add interest.

The circle of track, with its even 3% grades and 18″ radius curves, has been kept simple for a purpose. If the mountain turns out as well as you hoped, you'll be spending a lot of time just standing there watching the trains roll by. In such a situation, you don't want to worry about the train jumping the track on the other side. Either walk-around or stationary control may be used, depending on your preferences. The loop should be wired in the standard manner; see Paul Mallery's *Electrical Handbook, Vol. I.*

Trackplan IV is a synthesis of several elements. It is basically a double circle of track for people who like to watch trains run. However, it also provides quite a bit of switching and two dioramas. By necessity, the minimum radius is 15″. Although I've tried to squeeze in transition curves (easements), sectional track would easily fit. Usually the bane of all similar layouts is steep grades. Here the maximum has been held to 3%. This is accomplished by changing our thinking to suit the new layout concept. Usually when designing two return curves at the end of the trainboard where the two tracks are quite close together, it is common practice to place the higher one on the inside. It not only looks better, but also fits most scenery topology better. To do this on both ends of the table in this trackplan would require the tracks to cross over one another twice, one on each side. The necessary grades would be ridiculous. But, in actuality, we have a perfect solution: Keep the low track in the back, and hide it under scenery. This little bit of misdirection serves two purposes. First, it helps dispel the round-and-round look of the trackage. Second, it makes this diorama distinctly different from the one at the other end of the table.

I've crowded in quite a few sidings. Most come off level track to reduce difficulty in operation but are found at different heights to enhance the visual separation. Structures will have to be carefully chosen to maintain the effect. A small yard is provided for storage.

The result is a strange dichotomy. The plan as a whole sports lots of trackage with the center section of the table crowded to the nth degree. The two dioramas, on the other hand, provide space for uncluttered scenes. Seems like the best of both worlds to me.

Trackplan V was an attempt to create scenes within scenes. You can judge for yourself how successful it is. I've postulated a medium size logging company with several camps up in the mountains. The railroad weaves its way through the mountains, via a series of switchbacks, to reach each camp. Three of the camps are in one scene, as you can see. But each occupies a different height, faces a different direction, and exists in a separate physical location. To further distinquish the camps, none can be reached directly from one another. The loco must be worked through at least one other scene before reaching the next camp. Each camp has a different capacity for storage. Some are more difficult to work than others by nature of the trackage. Two require push service, two pull . . . altogether, an interesting piece of trackage.

Any trip will begin at headquarters, at the foot of the mountain. Here the company has provided some limited storage tracks for log buggies, a turntable and engine storage facilities. Several shanties provide office space and sleeping quarters. Further down-grade (perhaps on another module) would be the saw mill and more sidings. Only geared locomotives are used on this railroad because of the steep grades. Leaving headquarters, we start up-grade and into a tunnel. We emerge at Camp 1, pull onto the switchback, reverse direction and start upward again. Through another tunnel, the train then crosses a narrow valley (headquarters) and comes to a fork. We bypass the right turn which leads to Camp 2 and proceed upward at an even stiffer grade. We go under a long wooden trestle and pass Camp 3. The next switchback does not level off like the one at Camp 2. Here the process is to get a good downhill running start, to aid in getting to the top of the next grade. Having ridden on such a switchback at Rock of Ages Quarry at Barre, Vt., I can tell you it's an unforgettable experience—something like riding a roller coaster. At any rate, with the help of momentum our loco makes it up to Camp 4, where it exchanges empty buggies for full.

The concept of slow-moving trains going from scene to scene is ideal for walk-around control. Notice that in a tri-section like this, you can always follow the loco directly from one scene to the next; there are no blind spots. The only problem will be to establish a definite pattern of movement around the table to prevent snarls in the cord.

In trackplan VI, I've packed all the elements we've learned to use into one presentation. I call it "The Maze". Almost intuitively, the previous five plans developed the potentialities of the divider/backdrop, and I've led you through them as they occurred to me. For The Maze, however, I tried to formalize the rules and use them to ultimate advantage. First, space is gained by running parallel to the backdrop as much as possible. Second, visual distinction is achieved by gaining height *out of sight* and then bringing the track back into the scene. Third, running through the backdrop allows tracks to come and go at normally impossible directions. Fourth, the scenery can be totally different from scene to scene. Fifth, inventiveness is required to make the transference between scenes realistic. It helps to throw away all preconceived notions about what a layout should look like and how it should be designed.

The Maze has two long holding tracks

Six 4x8 trackplans to consider:

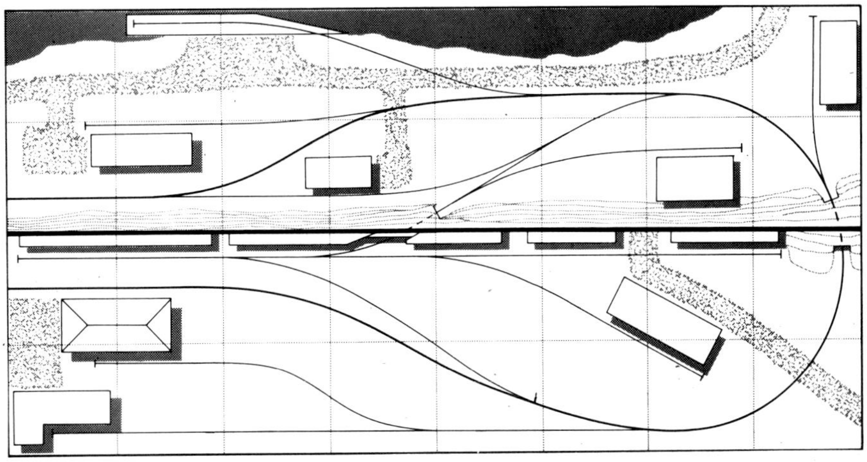

TRACKPLAN I

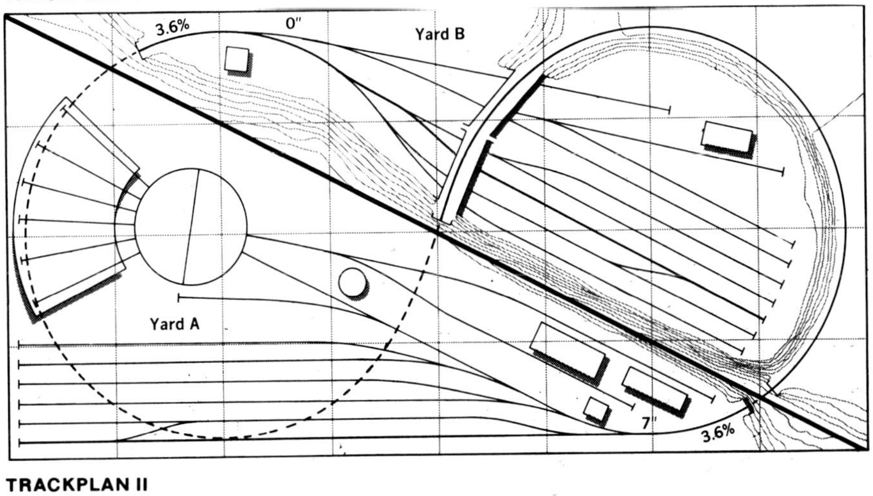

TRACKPLAN II

which serve as the setup track for the night's switching problem. A run-around allows the engineer to sort out the cars to some extent. Ten industrial sidings are provided, some of which can take more than one car. I suppose more sidings could be added, but we don't want to overly crowd things, do we? Logistics play a large part in escaping an operating session with some measure of sanity remaining. Several areas that serve as switchbacks are not very long, particularly if the track is already occupied. The direction sidings are entered following no special order, nor is the proper direction immediately apparent.

Some interesting ideas are possible in this layout, not ordinarily thought of. Since each scene should stand on its own, it behooves the operator to think along these lines. Stop assuming that the track continues to another, connected, scene. Run the main line through buildings if the main looks like a spur, as in scene X. So what if a track deadends just beyond the tunnel? You can still think of it as a through track and have trains originate there. It becomes obvious that the number of operating possibilities is limited only to your imagination.

The ordinary switching layout allows the operator to see all the tracks at once and to trace the most effective switching moves. The Maze will not allow this. In addition, as the main meanders from scene to scene, *almost no two consecutive spurs are in the same scene.* Therefore, the operator is forced to simulate the prototype. He must visualize the trackage scene by scene in his mind and determine his moves accordingly. As on some prototype roads, each day's work is an adventure.

Working the trains around the layout won't be much of a problem with walk-around control. The only track without a continuous vision transition occurs at the W-Y tunnel. Fig. 6 shows that the trackplan is, after all, relatively simple. The only unusual wiring required might occur at the crossing in scene X.

By happy chance, Modules D, E and F can be arranged to form a very interesting layout. I've just shown the basics of the plan in Fig. 7. With some very minor rearrangements, it provides a route for the lumber cut in the mountains of V, across and around VI, and finally to the sawmill in IV. A couple of other track connections will add further interest, as I've indicated in the plan. You might try your own arrangements to see if something better can be worked out. The area used here is approximately 12x14 feet, but requires a 16x20 room—not the most economical use of space, but then it wasn't designed to be. What it does show is how a room-sized layout could be constructed using the same principles as on a 4x8 board. I'm currently designing my latest layout along these lines.

By now, I hope I've convinced a few of you 4x8'ers that space is really not as big a problem as you thought. Remember, a piece of plywood with trains on it doesn't have to look like a piece of plywood with trains on it. I also hope that I've sold a few of you people with larger layouts on the general principles of limiting scope to specific scenes. That great scene *can* be yours!

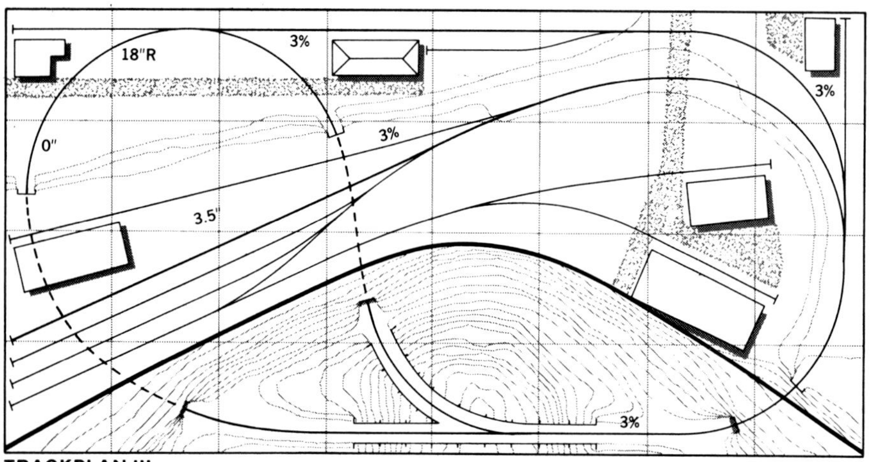

TRACKPLAN III

TRACKPLAN IV
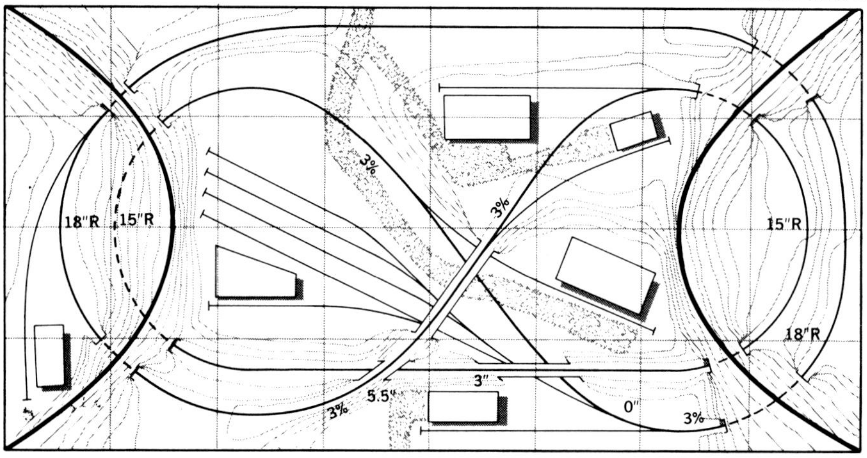

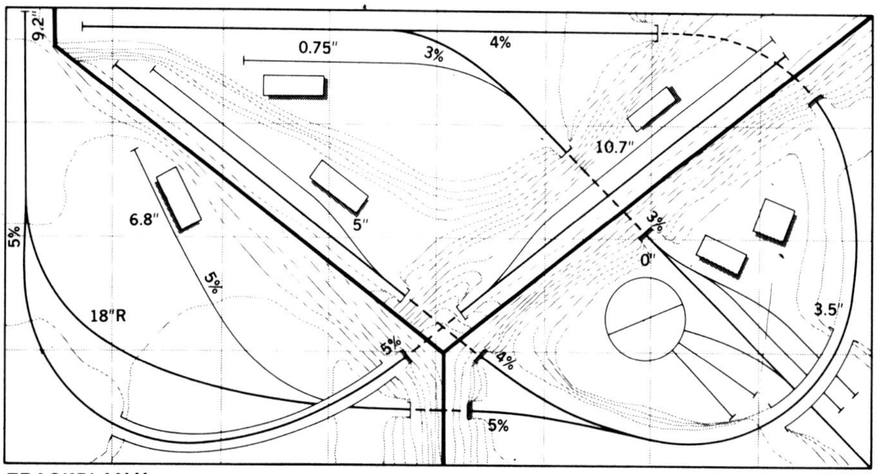

TRACKPLAN V

TRACKPLAN VI
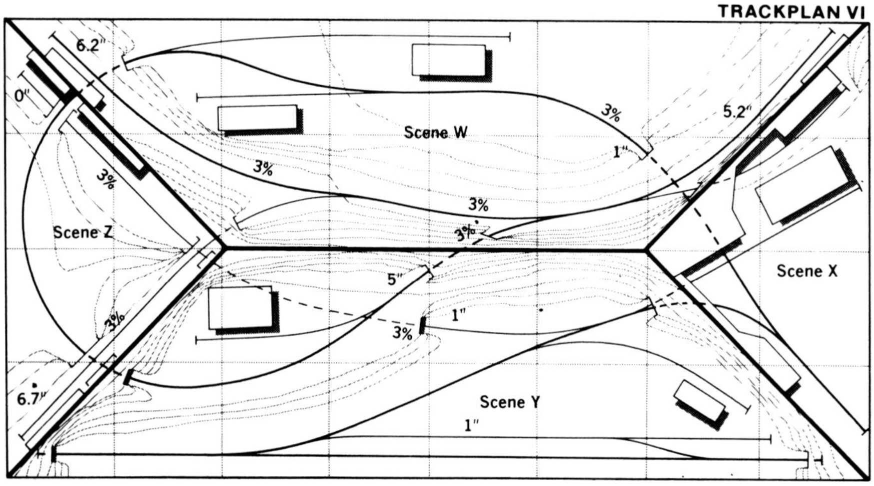

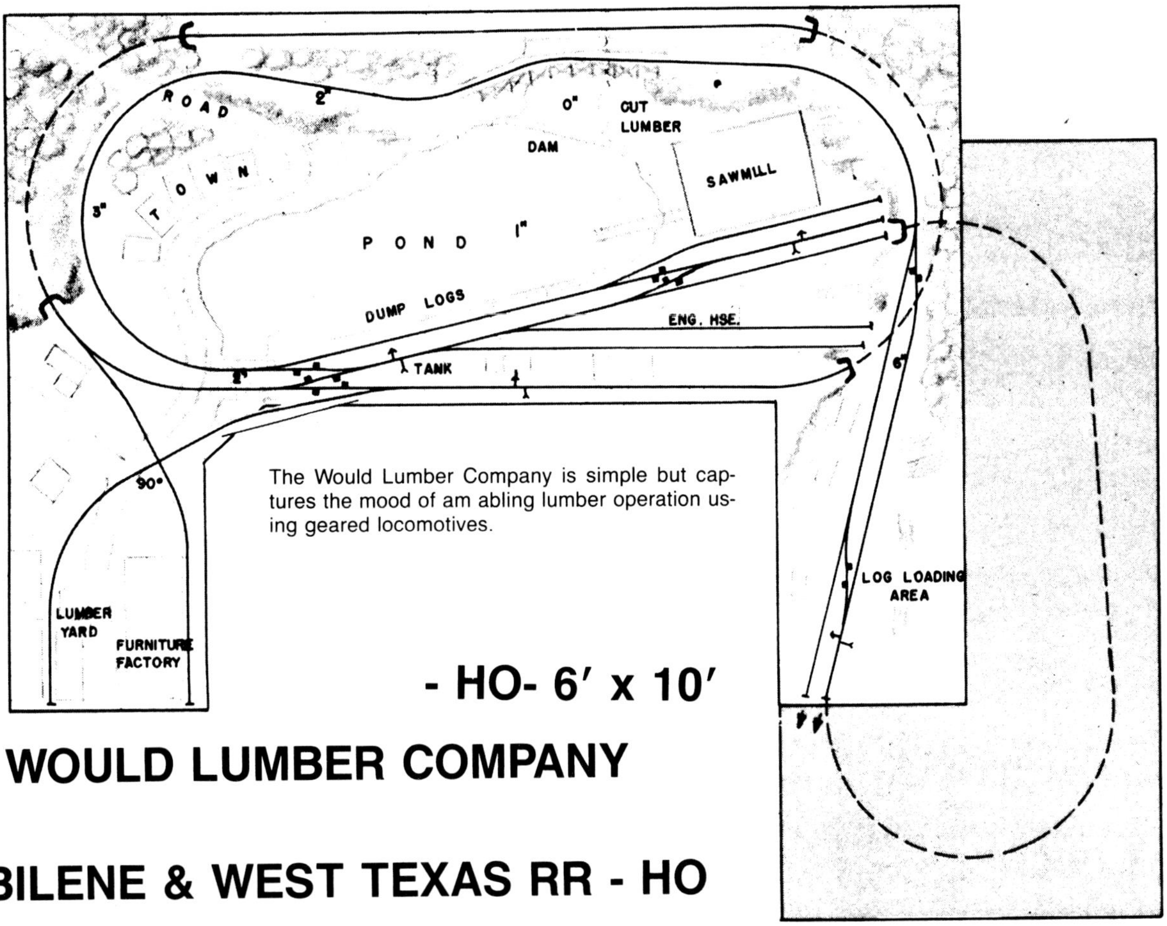

The Would Lumber Company is simple but captures the mood of am abling lumber operation using geared locomotives.

- HO- 6′ x 10′

WOULD LUMBER COMPANY

ABILENE & WEST TEXAS RR - HO

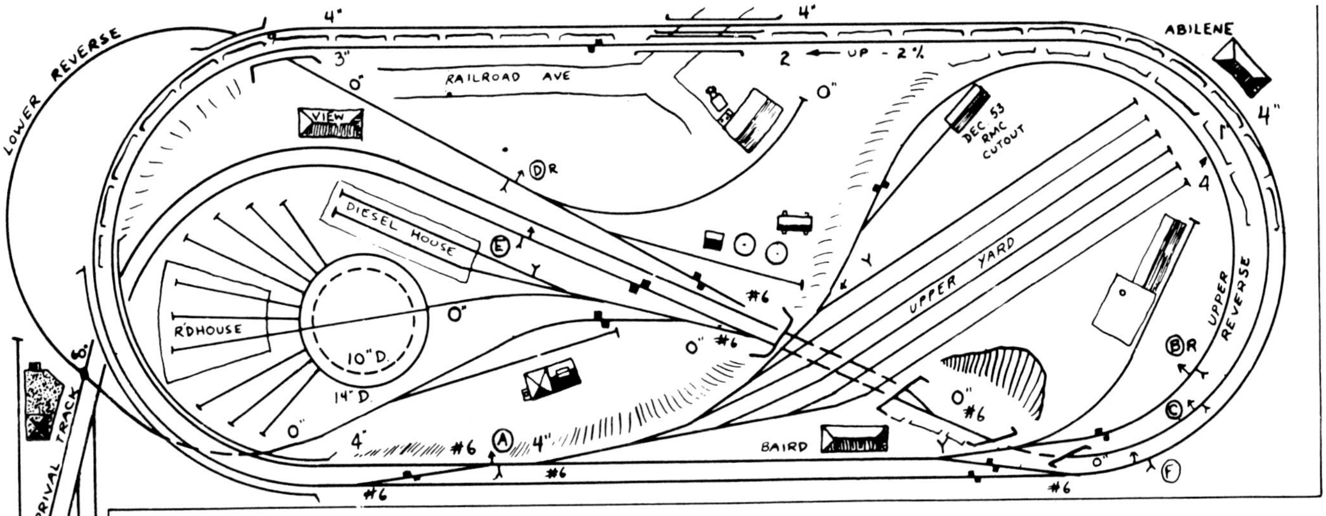

Insulated Rail Joiners (Gaps)

Power Feeders

Pack this layout design with long horned cattle and 1930 vintage stock cars or run it in a modern era with Amtrak and stack pack freights. This is Texas & Pacific country!

MOUNTAIN RAILROAD

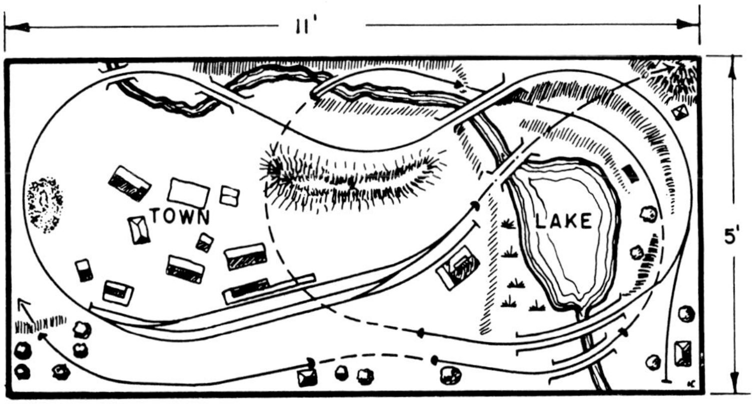

The Mountain RR is a simple point to point design which needs a simple terminal at each end or completion of loops at each end. Figure out the answer to fit your available space.

VERMONT EASTERN RAILWAY - HO

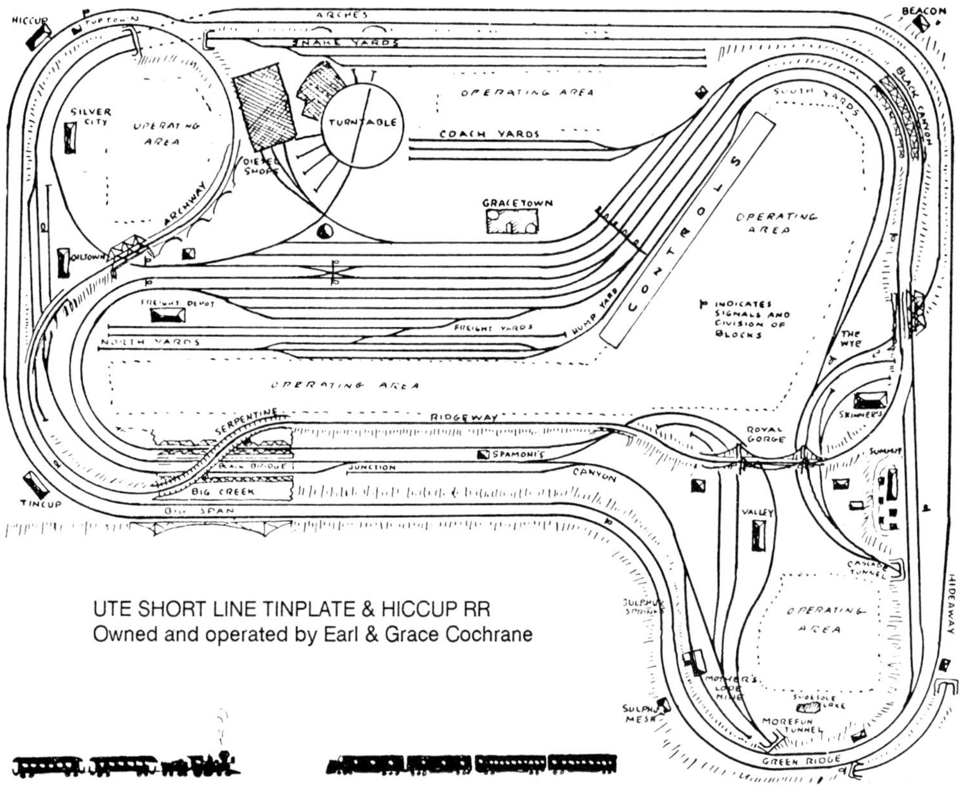

UTE SHORT LINE

HO

UTE SHORT LINE TINPLATE & HICCUP RR
Owned and operated by Earl & Grace Cochrane

CATSKILL MOUNTAIN RAILROAD - HO

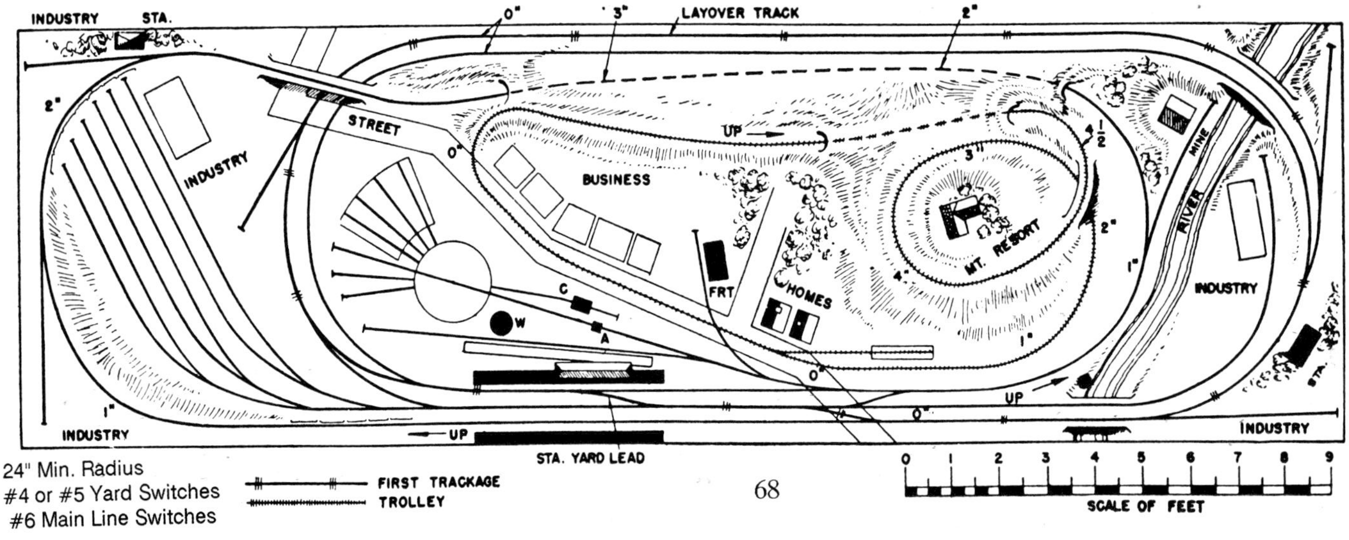

24" Min. Radius
#4 or #5 Yard Switches
#6 Main Line Switches

FIRST TRACKAGE
TROLLEY

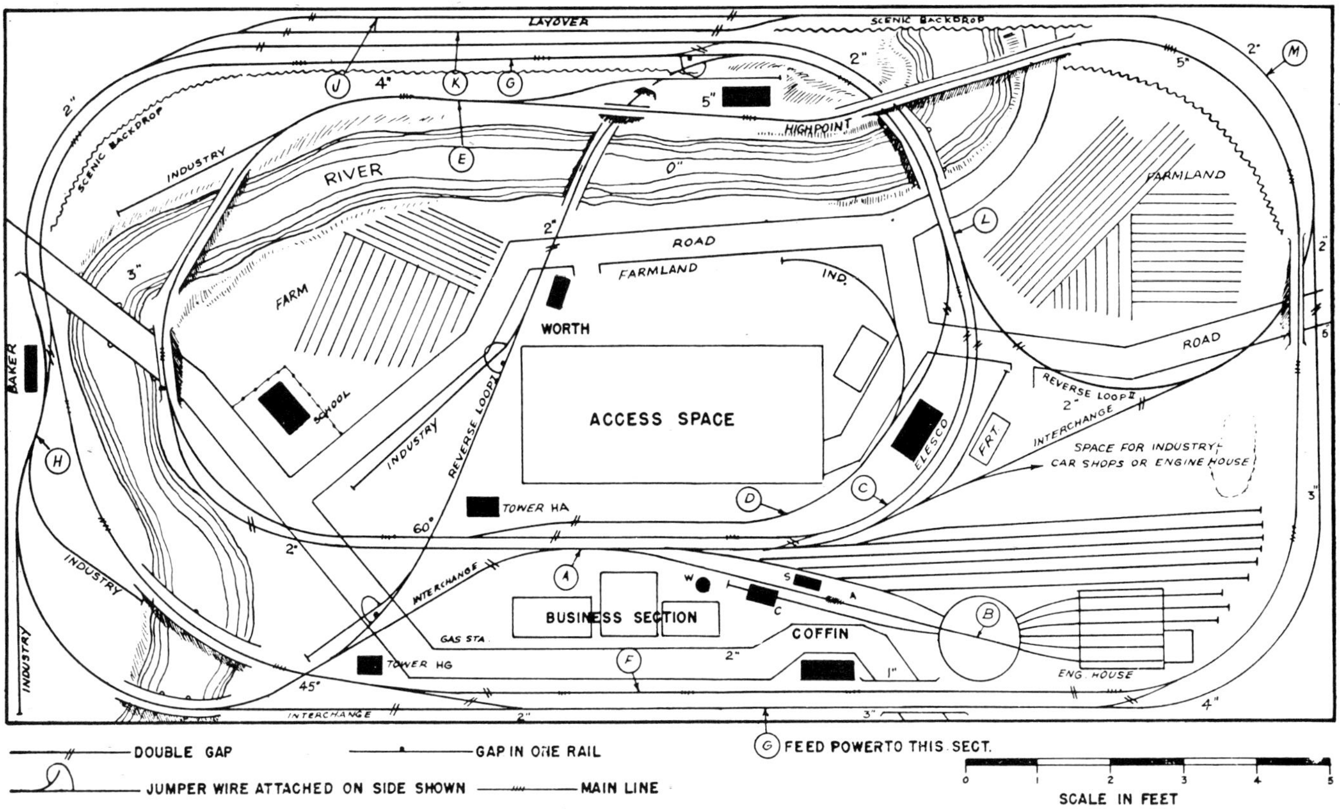

A large HO gauge railroad configured with an inside out figure eight.

HIGHPOINT RAILROAD - HO

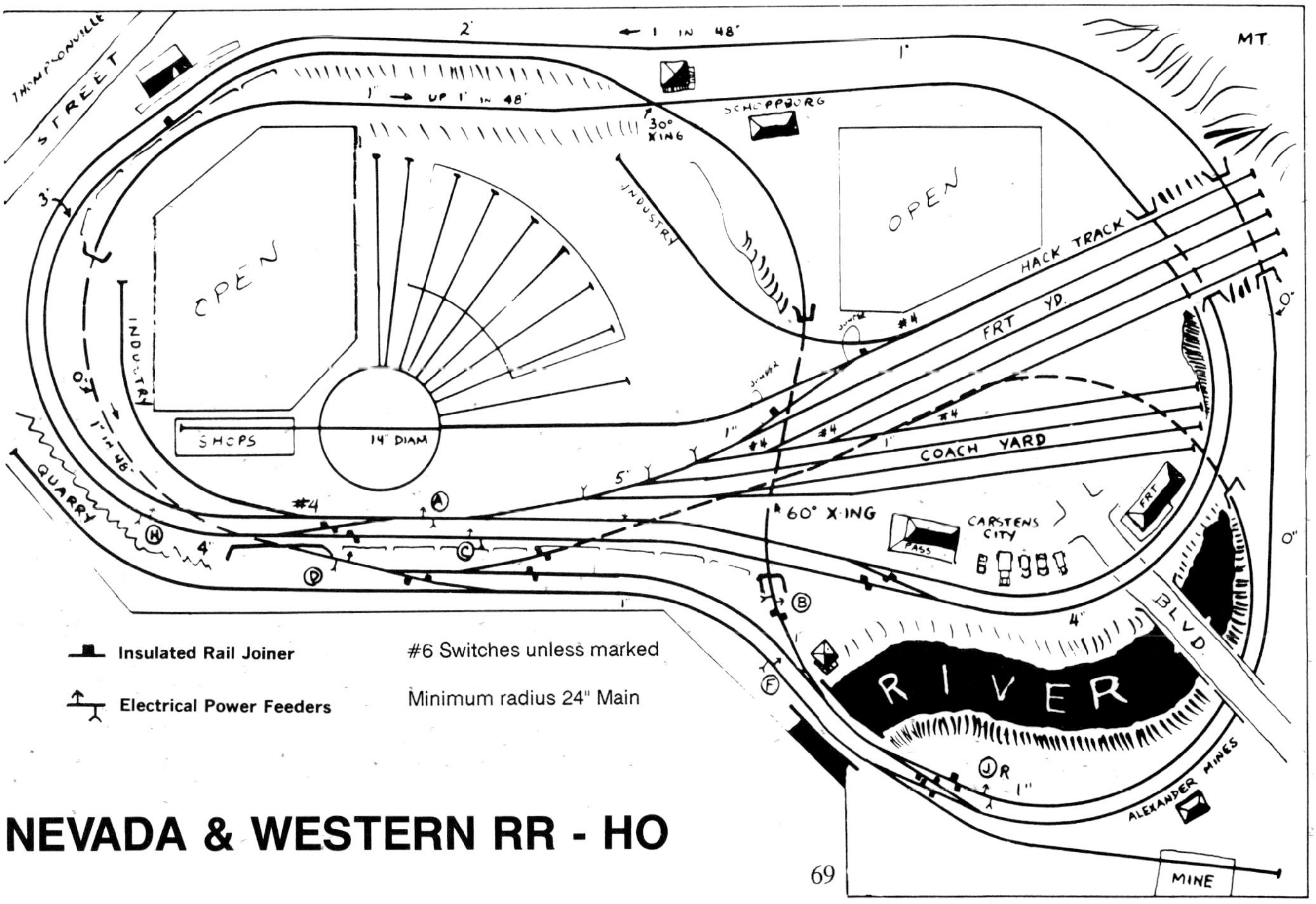

NEVADA & WESTERN RR - HO

BROOKLYN SOCIETY OF MODEL ENGINEERS HO - 18′ x 65′

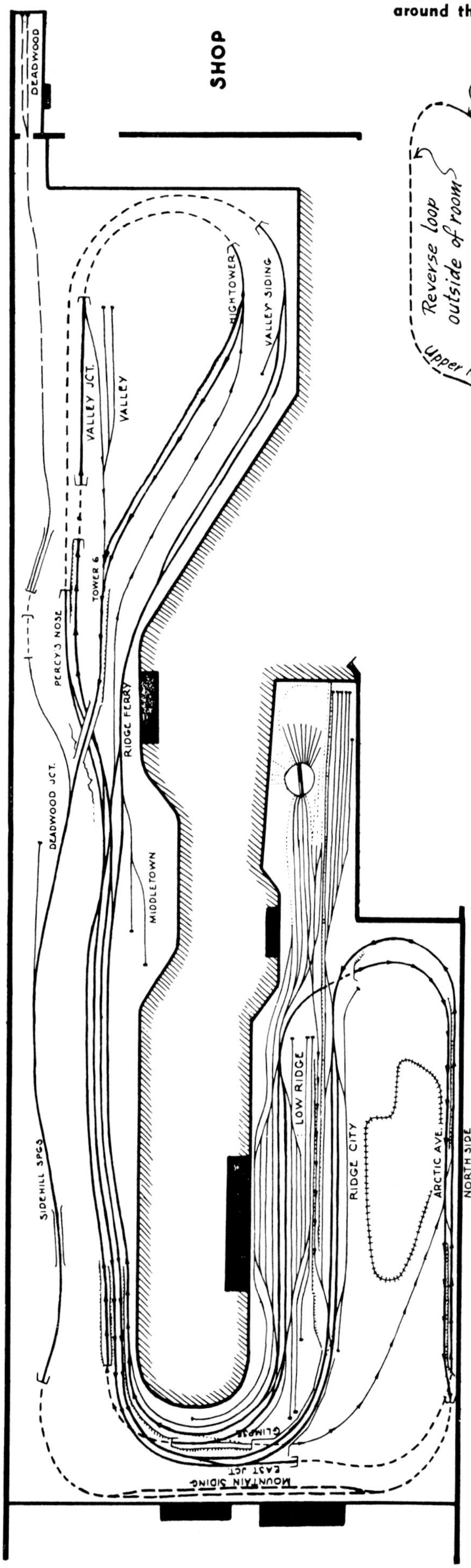

Layout for an HO gauge model railroad with overall dimensions 9½ x 11½ feet. Minimum radius on this pike is 22″, switches are all #6's. A single track railroad with one passing track and reverse loops at each end. The one track loops around the layout three times to an elevation of 11 inches.

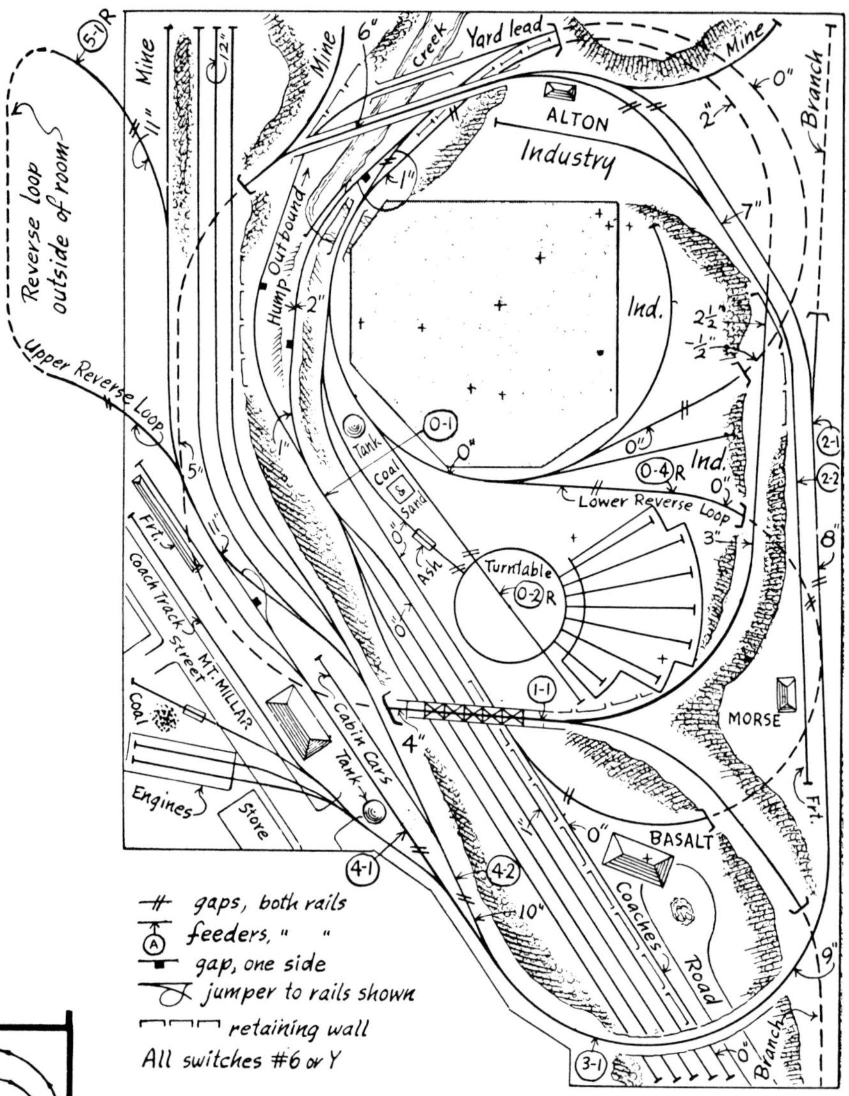

ST. LOUIS & ALTON RR - HO
HO – 9½′ x 11½′

IDAHO SOUTHERN RR - HO - 4′ x 8′

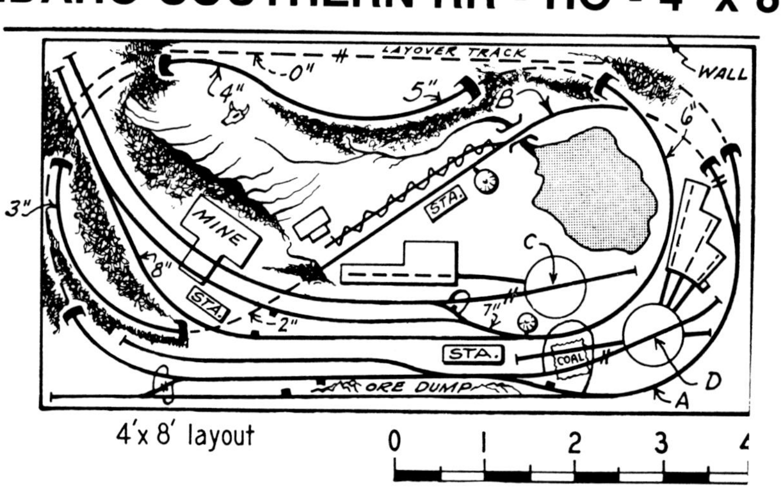

Station Locations

The Layout Doctor discusses prototypical station sites and how model depots may be placed on the layout.

As we ride along a railroad track, we seldom give thought to the locations of the passenger stations—even if they're no longer used. When the railroad was built through a town, the station was located convenient to the main street or, if there wasn't a town there, at some road crossing which soon became Main Street. Almost without exception, the location was on tangent (straight) track.

Why usually straight track? One reason immediately comes to mind: if a curved track is also used by non-stopping trains, it will be super-elevated (banked), making for awkward walking for passengers.

Model railroad stations require a little more thinking. Model railroads for the most part do not have an abundance of straight track. Many model pikes consist of two 180° curves connected by tangents of maximum length, which often isn't much. On top of that, the tangents (which are often two or three parallel tracks which you want to keep from looking parallel to themselves or to the walls or edge of table) are themselves snaked a little so that all of them are not really straight but gently and "realistically" curved.

Long straight stretches of track have the faculty of making a model railroad look longer. A station and sidings often make the straight stretch look shorter. It has been customary over the years to locate *model* stations on curves, especially the otherwise wasted triangular areas in the corners of the "typical" oval pike, or even in the purposely created dip in the otherwise long tangent.

It should be mentioned here that whereas real railroad stations are several miles apart, model railroad stations are for the most part quite close together, so that a train scarcely leaves A before it has to make a stop at B. A good way around this, if it bothers you, is to run your train one (or more) and a fraction times around the continuous oval (assuming your pike is so designed) between stops. That is, if you have four more or less equally spaced stations on an oval run, you start at A, make 1-1/4 circuits, and stop at B. Then 1-1/4 circuits more and stop at C, etc. Upon finally arriving at A again, you could then call it E and continue, stopping at F, G, H, then A again. You might use little tricks, such as where a "union depot" has a line on each side of it, each side representing a different stop. I assume you will do this manually with your controls, making the necessary bell, whistle, and other sounds at each stop. I've also done it automatically, working with a surplus solenoid operated 12 position rotary switch, so that besides the half dozen station stops I had, there was a long six times around "limited" run. Obviously, stops were a little abrupt with this simple control scheme.

Fig. 2a shows a typical model station location in the triangle outside an 18" radius HO curve. A real line normally wouldn't put a station here; if they did have to, they would not have the main line take the curve of the siding switch. Rather, they would have a switch on the nearby straight track so that the siding branched off, somewhat as in Fig. 2b. When the curve is sharp, the siding is short. (In Fig. 2a with an 18" curve, the siding has a useful length of about 12", enough for two short cars or one average car. On the other hand, a similar siding outside a 36" curve might have a useful length of 26"—four or five freight cars.)

Of what use is that siding, whether it's one or five cars long? If it is long enough, a freight station can be placed at its far end. If it's too short for that, the whole siding can be considered a team track, an open-air facility where the car can be unloaded directly into the wagons or trucks of an off-line customer. It is also possible on either long or short sidings to serve an industry of some type which may well be represented by scenery and a building "flat" against the wall. In locating such a corner siding, be very certain that (a) the switch is accessible for maintenance, and (b) if on a double-track line the switch points trail in relation to the normal direction of travel.

What about the platform; how do you make the curve correctly? Let me first explain to overseas readers that U.S. passenger platforms are normally low, so that steps on the cars are needed to go up and down. These low platforms are usually slightly above rail level. On a model railroad, my recommendation is that they be *at* rail level or a shade below. Why? Because one higher than rail level may cause derailments when hit by a rotating eccentric of a steam loco, or scraped by the low footboard of a diesel. Now, how do you make the curve of the platform follow the track? My way is to take a sheet of gray or black cardboard and lay it over the entire site. Then with your finger, run same heavily along over each rail. This will mark the cardboard with lines to cut on.

Putting the station on the inside of the curve as in Fig. 3a has several advantages. It frees the triangle outside the curve and its siding for "industrial development" (a freight station not crammed in, a factory building, a power house—anything requiring railroad freight service). It can also be filled with "scenery," either a hill with the railroad cutting through, or houses up on the bluff, these being especially suitable where a siding switch would be inaccessible. It also makes the railroad seemingly run *through* the scenery rather than just letting the scenery be an appendage to the railroad. If the station is at the back of your table, it will let you detail the back of your station instead of the front. Of course, you wouldn't have stations inside or outside *all* of the curves on your model railroad; all such scenery should be varied.

On double track you can let a station building on one side of the track serve both tracks by filling in the "devil strip" (as the trolley fans call it) between the double tracks. The close-in track is left open between the rails, except for one or two walkways at platform height opposite the door(s) of the station; see Fig. 3b. Trains do not pass *in* the station for safety reasons.

Another ruse on double track, if you have the space, is to have the main station on one side, usually on the side where most passengers will board the trains for the big city, and just a shelter on the other side with a platform for the few people who will wait for trains running the other direction. On straight track, it used to be customary for railroads to erect fences between the double tracks in such a situation to keep passengers from cutting across the two busy tracks. You could do it on curved track, too, if you spread the double track enough to give adequate clearance.

Where a siding, a secondary track, or even a yard lead has to pass between the station and its main track(s), the platform should be widened to include the lesser track, using the poster cardboard idea mentioned before, as in Fig. 4. However appealing it may sound, it is definitely not a good idea to make your station platform from plaster right on the spot. If your ties are wood or fiber, the wetness may warp them. If your ties are plastic, the wetness will not likely matter, but you will have one deuce of a time making changes (it will be almost impossible to thoroughly remove the plaster and have reusable track). Remember: *no layout is forever*; you will in time want to make changes.

Where there is no room on the layout at any of the logical spots for a passenger station, there are two alternatives. One, useful on urban type layouts, is to put the station *over* the tracks as in Fig. 5a, with stairways and maybe a baggage elevator leading to the platform(s). Fig. 5b shows what might be called "filled ground," with the station on a shelf extending out from the table. This would probably extend over your power pack on a shelf below, or any control facilities. You would not want to be forever catching yourself on it to the detriment of your clothing and the station.

You can locate a station on the table on straight track using the many prototype examples as a guide, but let me mention a couple of points about stations you might keep in mind. There is almost always a road crossing of some type at a station. Stores, hotels, banks, etc., often built around the depot, creating a city or village, though in a few instances there is just a road running to the station. The road will probably cross the tracks at grade, possibly requiring gates, blinkers, or at least a watchman and shanty. If there are switches (turnouts) at the station, try to route the road right through as many of the switches as you can. This makes constructing the crossing itself a little difficult, but if you use the poster cardboard rubbed over the rail and then cut out, it will work out quite well. If your switches have top-mounted switch motors, you will get some interesting highway/railroad crossing angles in avoiding these. Often times, these angles, and those 90° to them, will determine the angle of the gridiron of streets in the town behind the station.

At any station, you should have parking facilities, not just for the agent and crews deadheading to the terminal, but for passengers themselves and for people meeting trains. In the old days, there might well have been hitching posts.

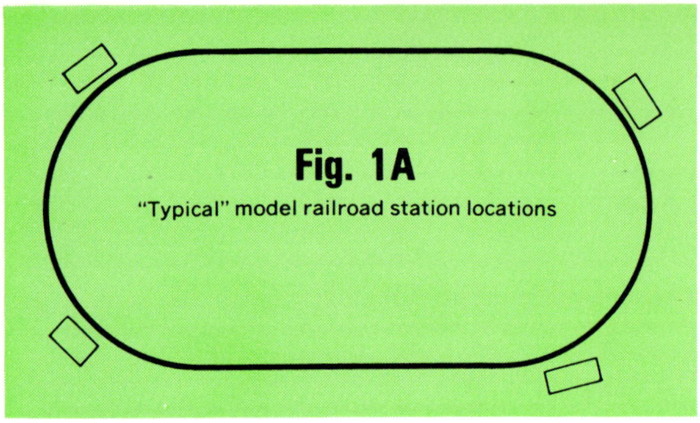

Fig. 1A
"Typical" model railroad station locations

Fig. 1B
Station in track "snake"

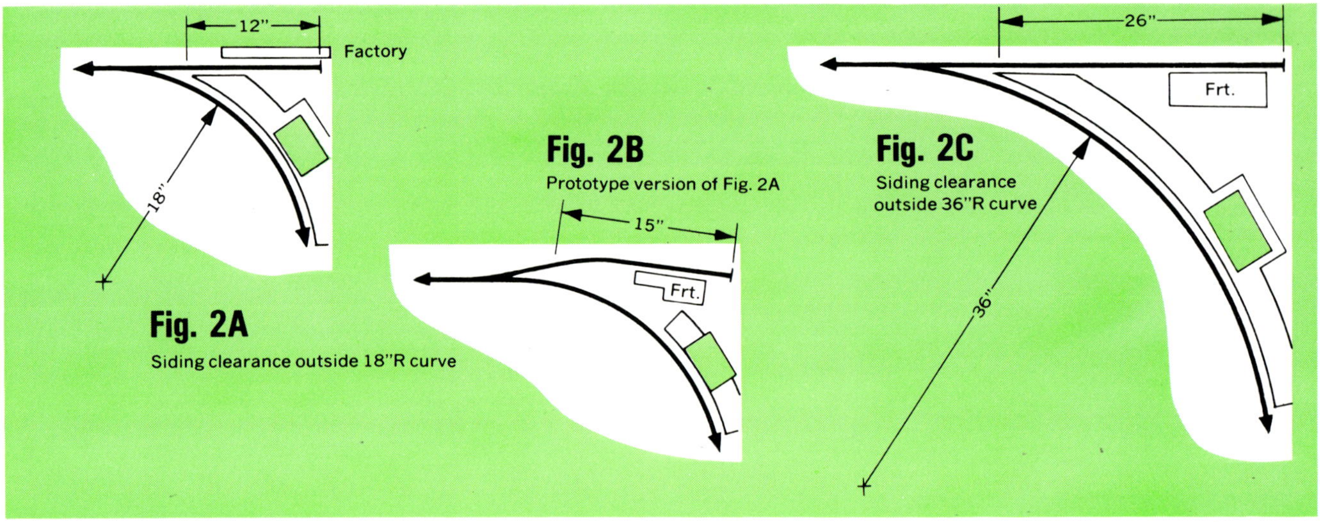

Fig. 2A
Siding clearance outside 18"R curve

Fig. 2B
Prototype version of Fig. 2A

Fig. 2C
Siding clearance outside 36"R curve

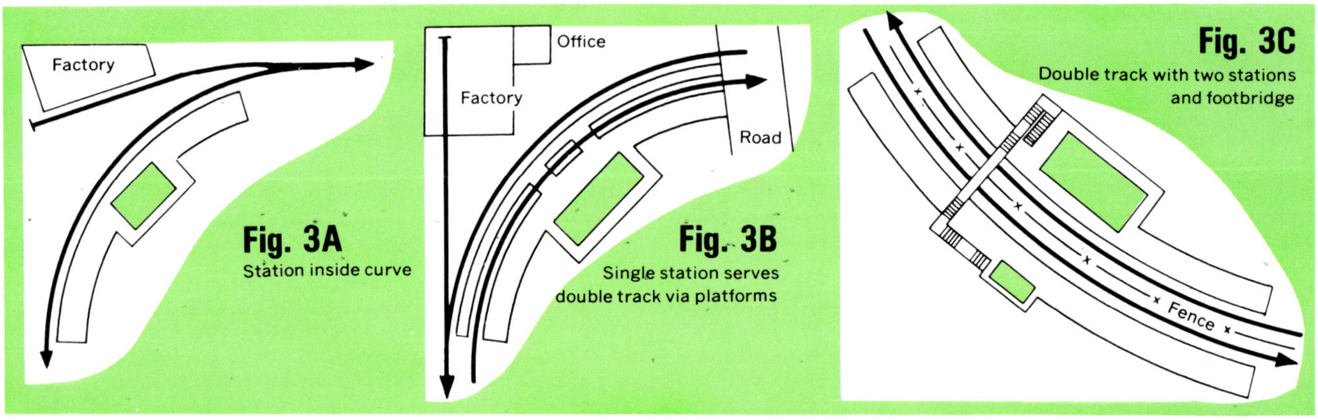

Fig. 3A
Station inside curve

Fig. 3B
Single station serves double track via platforms

Fig. 3C
Double track with two stations and footbridge

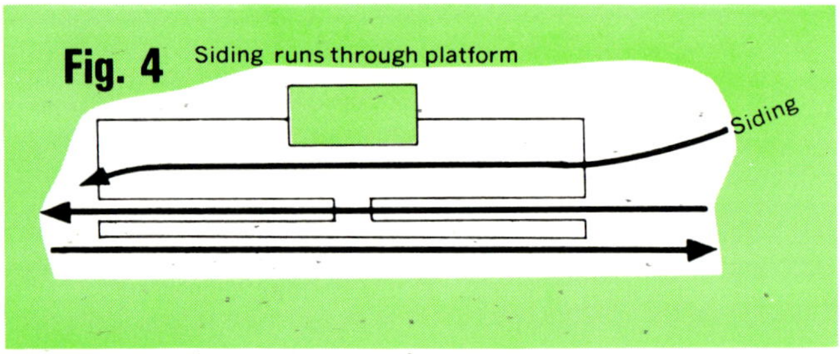

Fig. 4 Siding runs through platform

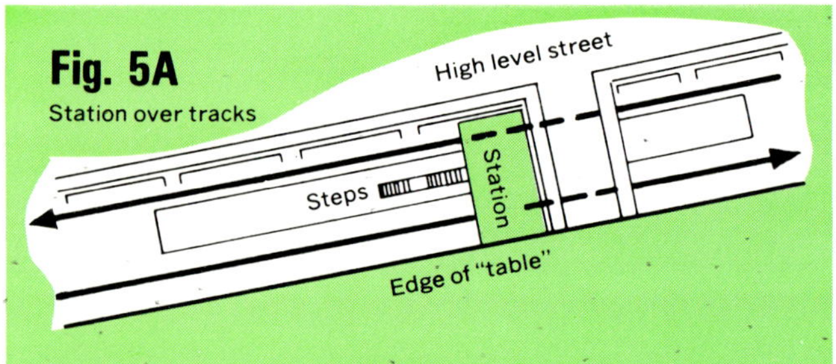

Fig. 5A
Station over tracks

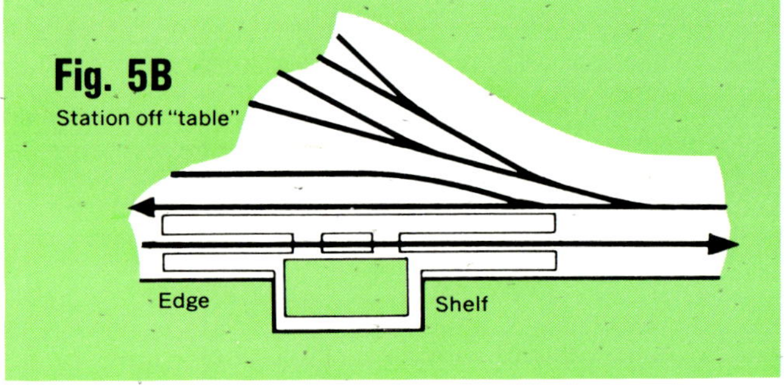

Fig. 5B
Station off "table"

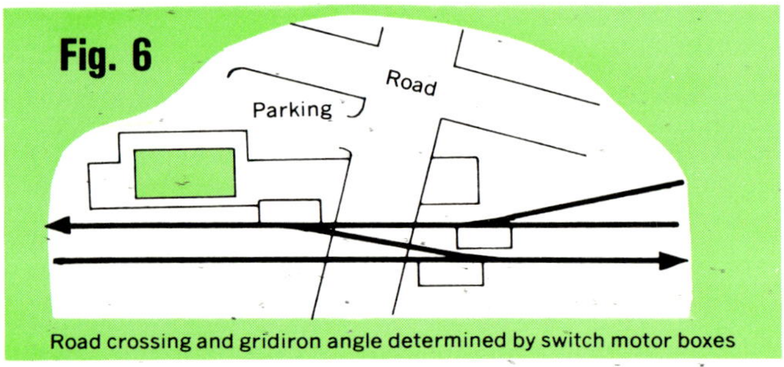

Fig. 6
Road crossing and gridiron angle determined by switch motor boxes